Analog Circuits and Signal Processing

The Analog Circuits and Signal Processing book series, formerly known as the Kluwer International Series in Engineering and Computer Science, is a high level academic and professional series publishing research on the design and applications of analog integrated circuits and signal processing circuits and systems. Typically per year we publish between 5–15 research monographs, professional books, handbooks, edited volumes and textbooks with worldwide distribution to engineers, researchers, educators, and libraries.

The book series promotes and expedites the dissemination of new research results and tutorial views in the analog field. There is an exciting and large volume of research activity in the field worldwide. Researchers are striving to bridge the gap between classical analog work and recent advances in very large scale integration (VLSI) technologies with improved analog capabilities. Analog VLSI has been recognized as a major technology for future information processing. Analog work is showing signs of dramatic changes with emphasis on interdisciplinary research efforts combining device/circuit/technology issues. Consequently, new design concepts, strategies and design tools are being unveiled.

Topics of interest include:

Analog Interface Circuits and Systems;

Data converters;

Active-RC, switched-capacitor and continuous-time integrated filters;

Mixed analog/digital VLSI;

Simulation and modeling, mixed-mode simulation;

Analog nonlinear and computational circuits and signal processing;

Analog Artificial Neural Networks/Artificial Intelligence;

Current-mode Signal Processing; Computer-Aided Design (CAD) tools;

Analog Design in emerging technologies (Scalable CMOS, BiCMOS, GaAs, heterojunction and floating gate technologies, etc.);

Analog Design for Test;

Integrated sensors and actuators; Analog Design Automation/Knowledge-based Systems; Analog VLSI cell libraries; Analog product development; RF Front ends, Wireless communications and Microwave Circuits;

Analog behavioral modeling, Analog HDL.

More information about this series at http://www.springer.com/series/7381

Jeffrey Prinzie • Michiel Steyaert • Paul Leroux

Radiation Hardened CMOS Integrated Circuits for Time-Based Signal Processing

Springer

Jeffrey Prinzie
KU Leuven
Geel, Belgium

Paul Leroux
KU Leuven
Geel, Belgium

Michiel Steyaert
KU Leuven
Heverlee, Belgium

ISSN 1872-082X ISSN 2197-1854 (electronic)
Analog Circuits and Signal Processing
ISBN 978-3-030-08745-6 ISBN 978-3-319-78616-2 (eBook)
https://doi.org/10.1007/978-3-319-78616-2

Printed on acid-free paper

This Springer imprint is published by the registered company Springer International Publishing AG part of Springer Nature.
The registered company address is: Gewerbestrasse 11, 6330 Cham, Switzerland

To my parents, who have always believed in me.

Preface

The goal of this research was to develop and test integrated CMOS circuits for radiation-tolerant time-based circuits with picosecond accuracy for nuclear applications and high-energy physics. The main applications for which these circuits were developed are time-based readout interfaces in high-energy physics particle detectors, clock generation, and data transmission for these detectors. During this research, a radiation-tolerant time-to-digital converter (TDC) and a low-noise clock synthesizer were designed and optimized for the particle detectors at CERN.

A short overview on the radiation effects and mitigation techniques to ionizing radiation is given, together with a discussion on the practical aspects which are required in modern TDCs and frequency synthesizers.

A high-resolution TDC is presented with a discussion on the design aspects and the practical implementations of the circuit which are required in nuclear environments. The TDC is based on a delay-locked loop (DLL) that has two phase detection circuits to boost the recovery time after an energetic particle disturbs the circuit. The functionality of the DLL ensures that the timing resolution of the TDC remains the same after irradiation. Furthermore, this DLL has a new phase detector architecture which reduces static-phase offsets in the phase detectors through a correlated sampling mechanism which has been implemented for the first time in the time domain. The circuit was prototyped in a 40 nm CMOS technology, and a 4.8 ps resolution was measured with a 4.2 mW power consumption.

DLL-based TDCs or serial communication links do require a low-noise, high-frequency reference clock. For a 64-channel TDC, a 2.56 GHz frequency synthesizer was designed to upconvert the 40 MHz reference clock of the Large Hadron Collider (LHC) at CERN to a 2.56 GHz high-speed clock with a targeted rms jitter below 1 ps. A radiation-hardened phase locked loop was designed in which both an LC-tank oscillator and a ring oscillator were present. The chip was prototyped in a 65 nm COS technology. These circuits were, in the next step, irradiated to make a comparison between ring and LC-tank oscillators in terms of noise, radiation damage, and single-event effects. The devices were irradiated with X-rays up to 600 Mrad to study the total ionizing dose effects on the circuits and were also irradiated with heavy ions to study the single-event effects on the oscillators. The clock generator has a power consumption of 11.7 mW and had an integrated rms jitter of only 345 fs. Triple modular redundancy was used in the digital circuits to protect them from soft errors. A new phase detector architecture is presented which minimizes the error rate due to high-energy particles in frequency synthesizers. The devices were also tested for temperature variations from −25 °C up to 125 °C.

From the results gathered in the radiation experiments, an improved LC-tank oscillator was designed which has a reduction of the sensitivity to single-event upsets of more than 600 times compared to a traditional implementation which is mainly due to the cross section of the tuning varactor of the oscillator. This technique was also experimentally verified.

Geel, Belgium — Jeffrey Prinzie
Heverlee, Belgium — Michiel Steyaert
Geel, Belgium — Paul Leroux

Contents

List of Abbreviations and Symbols

Abbreviations

ADC	Analog-to-Digital Converter
ADPLL	All-Digital PLL
AFC	Automatic Frequency Calibration
ASIC	Application Specific Integrated Circuit
BBPD	Bang-Bang Phase Detector
BW	bandwidth
CCO	Current Controlled Oscillator
CDR	Clock and Data Recovery
CML	Current Mode Logic
CTR	Counter
DCO	Digitally-Controlled Oscillator
DLL	Delay-Locked Loop
DNL	Differential Nonlinearity
DR	Dynamic Range
DT	Drift Tube
ELT	Enclosed Layout Transistor
FOM	Figure of Merit
FPGA	Field Programmable Gate Array
FSM	Finite State Machine
GRO	Gated Ring Oscillator
HL-LHC	High Luminosity LHC
INL	Integral Nonlinearity
ISF	Impulse Sensitive Function
LDD	Local Drain Diffusion
LET	Linear Energy Transfer
LF	Loop Filter
LHC	Large Hadron Collider
LIDAR	Light Detection and Ranging

LSB	Least Significant Bit
MBU	Multiple-Bit Upset
MDAC	Multiplying Digital-to-Analog Converter
MOMCAP	Metal Oxide Metal Capacitor
MOS	Metal Oxide Semiconductor
MOSCAP	Metal Oxide Semiconductor Capacitor
NIEL	Non Ionizing Energy Loss
NTF	Noise Transfer Function
OSR	Oversampling Ratio
PCB	Printed Circuit Board
PD	Phase Detector
PET	Positron Emmission Tomography
PFD	Phase Frequency Detector
PLL	Phase Locked Loop
PoR	Power on Reset
PSD	Power Spectral Density
PSRR	Power Supply Rejection Ratio
PWM	Pulse Width Modulation
RADAR	Radar Detection and Ranging
SEB	Single-Event Burnout
SEE	Single-Event Effect
SEGD	Single-Event Gate Damage
SEL	Single-Event Latchup
SET	Single-Event Transient
SEU	Single-Event Upset
SNR	Signal-to-Noise Converter
SoC	System-on-Chip
SPO	Static Phase Offset
SSB	Single-Sideband
STI	Shallow Trench Isolation
TID	Total Ionizing Dose
TDC	Time-to-Digital Converter
TMR	Triple Modular Redundancy
TOT	Time-over-Threshold
TPA	Two Photon Absorption
VCDL	Voltage Controlled Delay Line
VCO	Voltage Controlled Oscillator

Symbols

$< \cdot >$	Expectation value
$1/N$	Division ratio
A	Noise factor

C	Capacitance
c	Speed of light
c_n	Fourier series components
C_{ox}	MOS transistor oxide capacitance
Δf	Phase noise offset frequency
$\Delta\omega$	Phase noise offset pulsation
E	Energy
ϵ	Delay variation due to process variations
$\mathcal{F}\{\cdot\}$	Fourier transform
f	Frequency
f_s	Sampling frequency
$\Gamma(\omega_0 t)$	Impulse sensitive function (ISF)
G_m	Total transconductance
g_m	Transconductance
H_{cl}	Closed loop transfer function
H_{ol}	Open loop transfer function
$h(t)$	Impulse response
I	Current
I_{cp}	Charge-pump current
I_d	Drain current
K_{VCDL}	VCDL voltage-to-time gain
k	Boltzmann's constant
L	Inductance
$\mathcal{L}(\Delta f)$	SSB phase noise
$\omega_{-3\,\mathrm{dB}}$	Cutt-off frequency
ω_n	Natural frequency
ω	Pulsation
ω_0	Oscillator center pulsation
ω_p	Pole frequency
ω_z	Zere frequency
ϕ	Phase
$\phi_n(t)$	Time domain phase noise
P_{sig}	Signal voltage power
Q	Charge
Q	Quality factor
$q_{\max}$	Maximum charge swing on a node
R_{p-tot}	Equivalent parallel resistance
R_{ser}	Equivalent series resistance
R	Resistance
r_{ds}	MOS output resistance
$R_{\phi\phi}(t)$	Autocorrelation of the phase noise
σ_{cc}	Cycle-to-cycle jitter
σ	Standard deviation
$S_\phi(f)$	Frequency domain phase noise
T	Absolute temperature

T_{in}	TDC input time difference
τ_J	Instantaneous jitter
σ_{τ_J}	Standard deviation of the jitter
T_{LSB}	TDC bin width
T_q	TDC quantized output value
T_{ref}	Reference clock period
T_{ssp}	Single shot precision
$u(t)$	Unit step function
V	Voltage
V_{gst}	Overdrive voltage
V_t	Threshold voltage
$Z(s)$	Loop filter transfer function
ζ	Damping factor

List of Figures

List of Tables

Chapter 1
Radiation Effects in CMOS Technology

1.1 Radiation and Its Interaction with Matter

This chapter will introduce the radiation effects that are encountered in modern CMOS technologies that have been used in this work. A summary of the effects and the potential problems will be discussed.

To understand the damaging effects of radiation on silicon devices, a brief introduction on particle interactions is required. Radiation consists of sub-atomic particles or photons which may interact with matter as they travel through space. Significant research has been done, and is still being done, on how these particles interact with matter but the main mechanism will be listed briefly in this chapter. For CMOS technologies, the interaction of radiation with Si and SiO_2 is of main interest since the majority of the damage is observed in these materials of the devices.

Radiation can be sub-divided into two major categories, ionizing and non-ionizing radiation [1]. Ionizing radiation directly leads to ionization of the atoms of the matter which it passes through. This means that electrons are separated (temporarily) from the atom leading to free electrons and ionized atoms. Charged particles (like electrons, protons, ions, etc.) and photons can be involved in an ionization process. Non-ionizing radiation does not directly ionize the atoms and thus does not generate free charges as it passes through matter. Non-ionizing radiation are neutral particles (like neutrons and neutrinos) which have no charge and thus do not interact with the atoms' electrons. Neutrons do interact with matter through nuclear reactions between the neutron and the nuclei of the atoms. Reactions like "neutron capture" convert the matter's nuclei to different isotopes which may be unstable and decay to two lighter elements.

J. Prinzie et al., *Radiation Hardened CMOS Integrated Circuits for Time-Based Signal Processing*, Analog Circuits and Signal Processing,
https://doi.org/10.1007/978-3-319-78616-2_1

1.1.1 Direct Ionization

Ionizing radiation, excluding photons, can ionize atoms through coulomb forces between charged particles and the electrons of the target material [2]. Electric forces acting on the electrons may be strong enough to separate them from the atom. These electrons can become "free" electrons when the energy transferred to the electron is larger than the ionization energy of the atom. For silicon, this energy is 3.6 eV and is the energy required to bring an electron from the valence band in the conduction band.

Figure 1.1 shows an example of an ionization of an electron in a silicon bandgap. The bandgap of silicon is 1.1 eV but since silicon is a non-direct semiconductor, the ionization happens through a phonon generation in the lattice. To allow this, an energy of 3.6 eV is required which is partially transferred to the electron and phonon to ensure a conservation of energy and momentum.

Coulomb forces between charged particles and the atoms' electrons can happen as is shown in Fig. 1.2. These effects are called direct ionization. A charged particle can pull or push to the electron when being in its neighborhood. With sufficient force, the electron can be separated from the atom leading to an ionized atom. Positive particles can cause direct coulomb collisions with the electrons pushing them out of the atom's orbitals. If charged particles like electrons and positrons travel in a potential field, the electric forces will bend the trajectory of the particle leading to additionally generated photons, more known as "bremsstrahlung" which may result in secondary ionization effects.

Heavy ions ionize in a similar way but can also cause collisions with the nuclei giving nuclear reactions resulting in lighter ionizing particles which again lead to ionization.

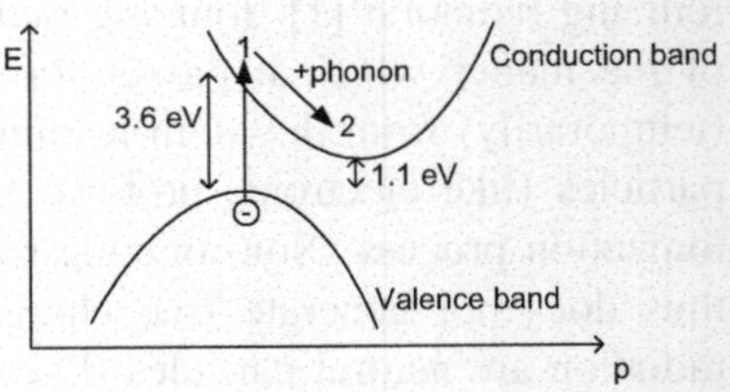

Fig. 1.1 Electron ionization in indirect bandgap semiconductors requiring more energy than the bandgap energy. The graph shows the energy bands as a function of the electron momentum

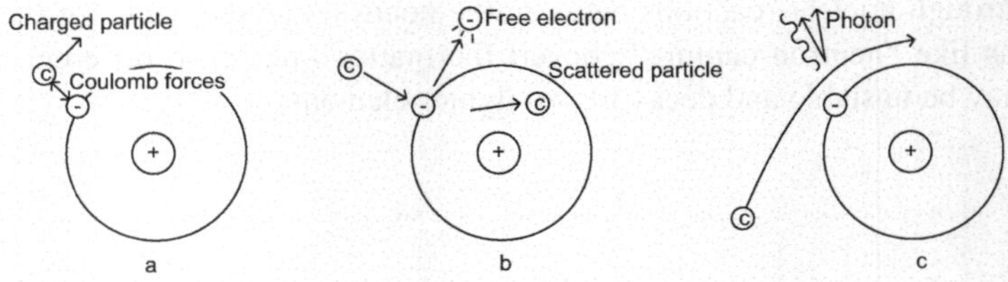

Fig. 1.2 (**a**) Coulomb force interaction (**b**) Coulomb collision (**c**) Bremsstrahlung

1.1.2 Electromagnetic Radiation

Photons are a special case of ionizing radiation [3]. X-rays and γ rays consist of highly energetic photons. Since photons have no charge and mass, they cannot interact with matter in the ways discussed previously. There exist no coulomb forces between the photon and the electrons. Photons interact with matter through the photoelectric effect, Compton scattering, or pair production as is shown in Fig. 1.3 [4]. Even though the mechanisms are different compared to coulomb forces, the result is the same, free electrons. Einstein's theory posed that free electrons are observed when a cathode is irradiated with photonic-radiation. The energy and momentum of the photon is absorbed by the electron and the atom and the excess energy above the ionization energy is converted into kinetic energy of the electron. Compton scattering is similar to the photoelectric effect in which the photon is not completely absorbed. The photon interacts with the target electron transferring part of its energy and momentum to the electron, again leading to free electrons when the energy is sufficient. Finally, free electrons can be generated when no electrons exist. This effect is called pair production in which an electron and a positron are generated from a photon. This requires a minimum energy for the interaction to occur since the rest mass of both particles requires a minimal energy to be created ($E = m_{rest}c^2$, where m_{rest} is the combined rest mass of the electron and the positron). The minimum energy for this event is 1.02 MeV. Additionally, the excess energy transferred to the electron should be sufficient to be a free electron. This reaction only occurs when the photon is at an interaction point, like nucleon and does not happen in free space since it requires a conservation of momentum.

In Fig. 1.4, the probability is shown for these three effects to occur. The photoelectric effect occurs at low photon energies. The probability reduces when the photon energy is higher since the momentum of the photon has to be transferred to the electron. At moderate energies, Compton scattering is therefore more likely. For higher energies, the photons are absorbed through pair production which dominates Compton scattering [4].

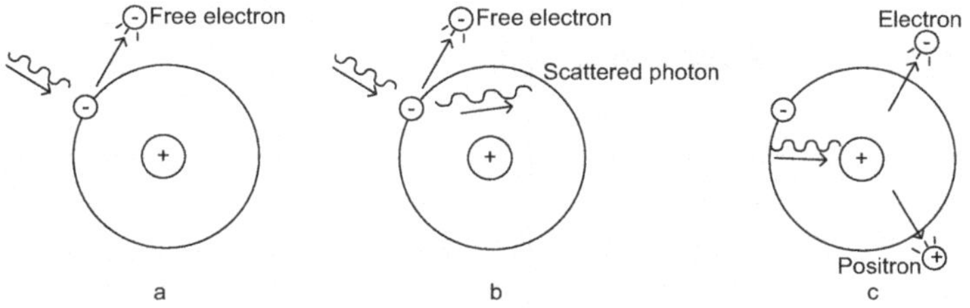

Fig. 1.3 (**a**) Photoelectric effect (**b**) Compton scattering (**c**) Pair production

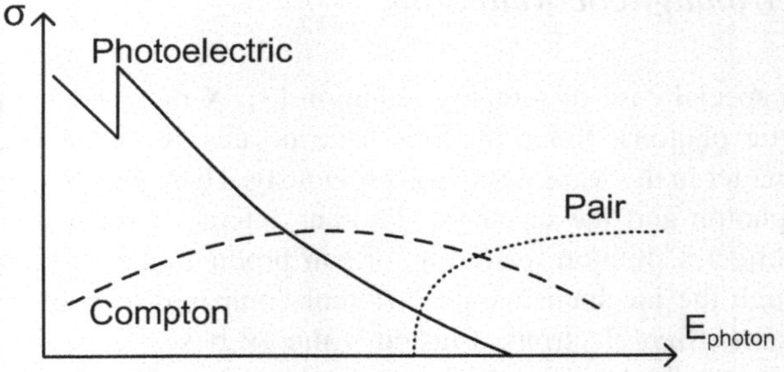

Fig. 1.4 Cross sections (probability) of the photon interactions as a function of photon energy

1.1.3 Neutrons

When neutrons travel through matter, they may interact with the nuclei of the atoms and not directly with the electrons. When a neutron is in the range of the target nuclei, it can be captured by the nuclei [5]. This effect is called neutron capture. Since neutrons are not charged, they are not repelled by the atom. An additional neutron can make the nuclei unstable and decays to smaller fragments. The target nuclei typically decay to smaller elements with an additional α-particle and photon. These secondary particles, generated from the nuclear reaction can lead to ionization referred to as indirect ionization since the neutrons do not directly ionize the atoms. Since the neutron has to directly interact with the nucleus and has to destabilize the nucleus, it takes some time for the reaction to occur. This is the reason why slow (low energy neutrons) have higher cross section (the spatial probability for an interaction to occur) compared to fast (high energy) neutrons since the potential interaction time of slow neutrons is higher than fast neutrons. A big portion of the energy is not converted into free electrons. This energy is called Non Ionizing Energy Loss (NIEL).

1.1.4 Effects on Semiconductors

The above discussion briefly introduces the basic interactions of charged particles, photons and neutrons with matter leading to ionized atoms and free electrons. To discuss radiation effects in CMOS devices, generated electrons are important. Note that, when a free electron is generated from a stable atom within a semiconductor, a hole is also generated in the ionized atom. Therefore, it can be said that nuclear radiation generates electron-hole pairs in the semiconductor such as silicon. The same holds true for SiO2 with the difference of a larger band gap.

The generation of electron-hole pairs is the first step in the radiation damage in semiconductors. In the second step, these charges change the electric characteristics of the devices. These effects are discussed in Sect. 1.2. Furthermore, these free electrons lead to transient currents after they are generated inside the transistors of the circuits. These effects and mechanisms are discussed in Sect. 1.3. Radiation effects require additional simulations to predict the behavior of the circuit after or during irradiation. This has become difficult in modern technologies. A simulation flow is introduced in Sect. 1.4.

1.2 Total Ionizing Dose Effects

Total Ionizing Dose (TID) effects on CMOS technologies originate from trapped charges in the oxides around the transistors. When ionizing radiation passes through the transistors' oxides, SiO_2 atoms are ionized and electron-hole pairs are generated. In the old technology nodes, from 0.13 μm and above, the majority of the radiation effects were seen in the gate oxide where charges were trapped. While many circuits resort to technologies with smaller feature size, the basic charge trapping mechanisms remain valid in deep submicron CMOS technologies.

1.2.1 Basic Charge Trapping in CMOS Transistors

The basic charge trapping mechanism is shown in Fig. 1.5 which shows the energy band of the gate, oxide, and silicon interface [6]. The charge trapping happens in three phases. In the first phase, when an ionizing particle crosses the oxide, electron-hole pairs are generated inside the oxide. As discussed before, one particle may generate multiple pairs depending on its total energy and interaction with the oxide. After the atoms are ionized, a fraction of these free electrons will immediately recombine without creating any damaging effects in the transistors. The remaining free electrons which do not recombine are relatively mobile in the oxide and are collected by the gate node under positive bias (nmos). This results in positively charged holes left in the oxide. In the second phase, these positive charges migrate in the oxide through localized states towards the silicon interface [6]. The hopping mechanism has been known to be thermally and electric field activated. Once they finally arrive near the silicon interface, the charges can be trapped and remain present in the device.

Since the trapped charges are positive [7, 8], they change the threshold voltage due to oxide traps (V_{ot}) of the devices by

$$V_{ot} = -Q_{ot}/C_{ox} \qquad (1.1)$$

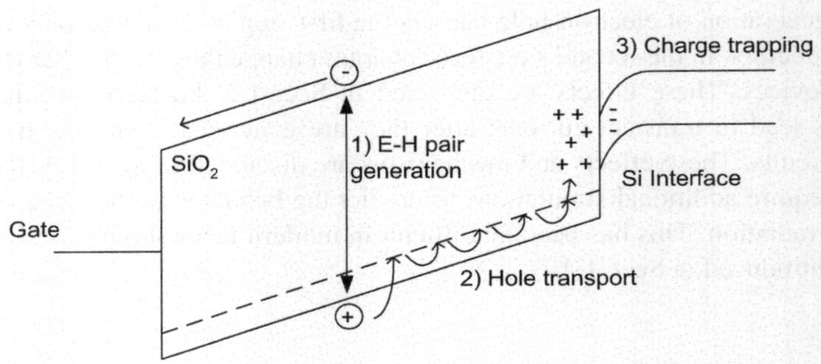

Fig. 1.5 Band energy representation of the gate-oxide-silicon interface and the mechanism of positive charge migration and trapping

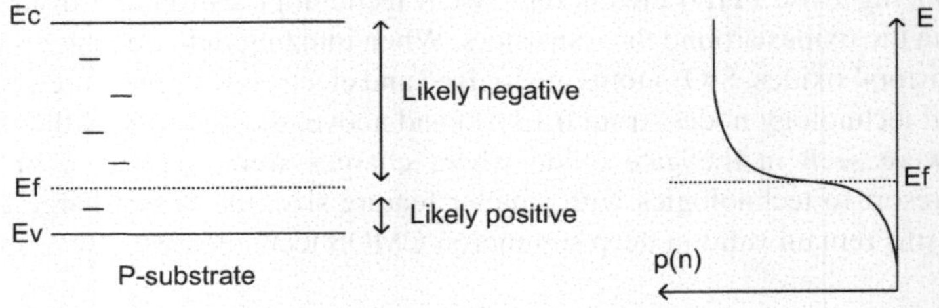

Fig. 1.6 Bandgap of a P-type substrate and probability of charges in the traps within the bandgap. E_f, E_c, and E_v are the fermi, conduction band, and valence band energy, respectively

The amount of charge which is generated by the radiation is directly proportional to the gate thickness since a thicker oxide has more yield in generating charges. Therefore, the threshold voltage shift is directly proportional to t_{ox}^2.

$$V_{ot} \propto t_{ox}^2 \tag{1.2}$$

From this result it can be concluded that scaled technologies are advantageous for TID effects since thinner gates capture fewer charges and have higher C_{ox} [9].

Besides, charges generated in the oxide, the radiation damage leads to a buildup of traps near the interface of the SiO_2 [10, 11]. These traps can be neutral, donor or acceptor type. In nmos transistors, which are fabricated on a p-substrate, the fermi potential is below the midband energy. Therefore, energy levels within the bandgap are likely to be trapped by negative charges. For pmos devices, these traps are occupied by positive charges since the fermi potential is above the midband energy level. Figure 1.6 shows an energy band diagram of a P-type substrate (nmos transistor) with the probability of a negative charge at a given energy level. These probabilities follow the Boltzmann distributions of excess carriers in the semiconductor [12].

While oxide trapped charges are positive for both nmos and pmos transistors, trapped charges due to radiation induced traps are positive for pmos devices but negative for pmos devices [13].

The overall threshold shift can be calculated as the sum of oxide traps (Q_{ot}) and interface traps (Q_{it}) for P− and nmos devices

$$V_{tot} = V_{ot} + V_{it} = -\frac{Q_{ot} + Q_{it}}{C_{ox}} \tag{1.3}$$

For pmos transistors, both Q_{ot} and Q_{it} are positive such that both effects lead to a negative shift of the threshold voltage. In nmos devices, oxide traps are positive while interface traps are negative leading to a competing effect. Since interface traps are observed at a later stage after irradiation, typically a reduction of the threshold voltage is observed in the first phase while later, the threshold voltage increases again.

The above results discussed suggest that TID effects are uniform on all devices on the chip. This is true if the irradiation gradient is zero (which can be assumed for small chips) and the devices are identical. The latter is not the case since local process variations make each transistor unique. In [14], experiments were done on the mismatch between CMOS transistors before and after irradiation from which it has been found that the variability increases with the dose which is likely due to the impact of random dopant fluctuations on TID effects.

Finally, leakage currents increase as well in nmos transistors due to the reduction of the threshold voltage. This can become a dramatic concern in large digital chips since the power grid may become insufficient [15].

1.2.2 Narrow Channel Transistors

The mechanisms described above are originating in the gate oxide and were the major contributions in old technologies. However, as transistors shrunk in both length and width, secondary effects started to occur in other relevant oxides in the transistor [16]. A narrow transistor is a transistor which has relatively large length but small width.

Figure 1.7 shows the layout of a narrow nmos transistor and its cross section. To define the width of the channel, STI (Shallow Trench Isolation) is used which is SiO_2 oxide. As the gate oxide can trap positive charges, so do STI oxides. The STI however is significantly wider compared to the gate oxide and can trap much more charges than the gate oxide. However, its effect is only seen near the edges of the transistor [17]. Therefore, the STI oxide traps become significant when the width of the transistor is shrunk such that the edge effects of the trapped charges become relevant and change the behavior [18]. Wider devices are less influenced by the STI traps since the channel potential disturbance only happens near the edges.

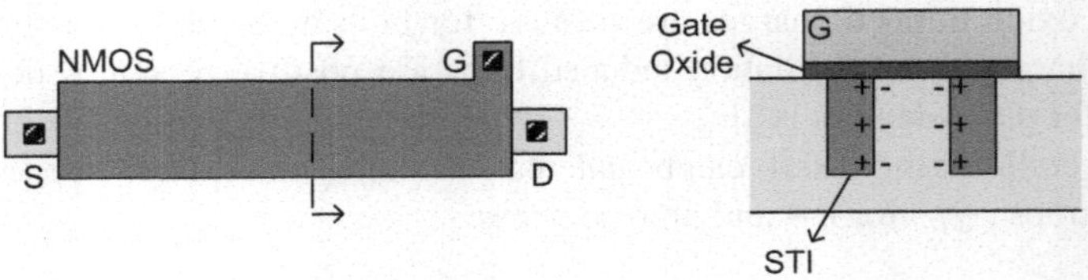

Fig. 1.7 Narrow nmos transistor and its cross section. Positive trapped charges disturb the local potential in the channel near the edges of the transistor

Fig. 1.8 Positive oxide traps below the LDD spacers invert the p− implants which leads to an increased series resistance

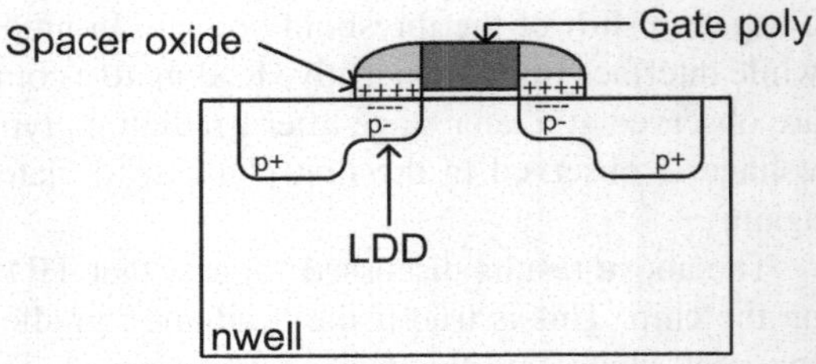

1.2.3 Short Channel Transistors

In a short channel device, for which the length is small, a different and significant important effect is observed from TID measurements which is disastrous for pmos devices. Figure 1.8 shows a cross section of a short pmos transistor. To reduce the electric fields in the channel, LDD (Local Drain Diffusion) is used in modern devices. This reduces hot carrier injection to improve the reliability of the devices. The LDDs are fabricated through spacers at the edges of the poly gates which prevent highly doped source-drain implants below the spacers. These spacers are associated with a relatively thin (15 nm) oxide. For pmos devices, this results in locally low-doped p− extends of the highly doped p+ source and drain.

Similarly to the gate oxide and STIs, this oxide below the spacer can also trap positive charges. As a result, the local effective doping of the p− LDD is further reduced since the positive charges in the oxides below the spacers influence the potential in the LDD. This leads to a tremendous increase of the resistance of the LDD spacers [16].

The effect is happening for all different lengths of the pmos devices but becomes visible as the length becomes shorter and the ratio between the LDD resistance and the channel conductivity becomes larger.

This effect is seen in pmos devices since the implants are positive. For nmos transistors, the n− LDD implants become stronger. It is believed that this reduces the effectiveness of the LDD and increases the hot carrier injection [19, 20]. Measurement results have also shown that the degradation is highly asymmetrical showing that the main degradation is happening near the drain of the nmos devices as a result of increasing hot carriers [16, 21].

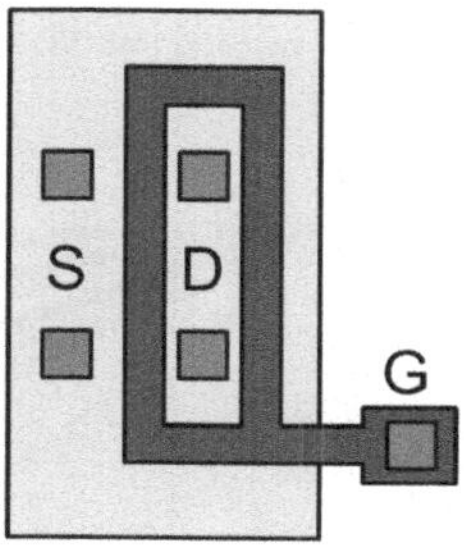

Fig. 1.9 Enclosed layout transistor to mitigate edge effects in narrow transistors

1.2.4 Enclosed Layout Transistors

The effect of STI can be eliminated by the use of Enclosed Layout Transistors (ELTs) [22]. In an ELT, as is shown in Fig. 1.9, the drain or source is fully enclosed by a gate which excludes any edges in the transistor and avoids the effect of those charges. To meet the design rules of the technology, the minimum width of an ELT is typically much larger compared to a minimum sized linear transistor which leads to a significant increase of the power consumption in digital circuits which employ standard cell libraries that are designed with ELTs [23]. However, as the sizes of the transistors increase, the radiation hardness is significantly improved. An experiment was performed in [24] to investigate the degradation of the speed of digital cells. Since the speed is proportional to the drive current, the cells become slower with increasing dose. Figure 1.10 [24] shows the speed reduction for different standard cell libraries in the same technology. It can be seen that small cells (7 track) are significantly more sensitive to TID compared to larger cells. Clearly the ELT outperforms all libraries since it eliminates the edge effects in the transistors. However, the 18 track ELT library contains the smallest transistors that could be used to comply with the foundry DRC rules.

1.2.5 Experimental Results

Experiments were performed in [16] to investigate the radiation hardness of a 65 nm CMOS process. During these experiments, devices with different geometries were measured. Figure 1.11 [16] shows the current through a pmos device which shows the effect of narrow and short devices. Narrow transistors degrade through STI trapped charges while short transistors degrade through LDD resistance. Therefore, the combination of short and narrow devices is the worst. It is clear that the ELTs perform best in the experiment since no STI effects occur. The remaining degradation originates in the LDD short channel effects.

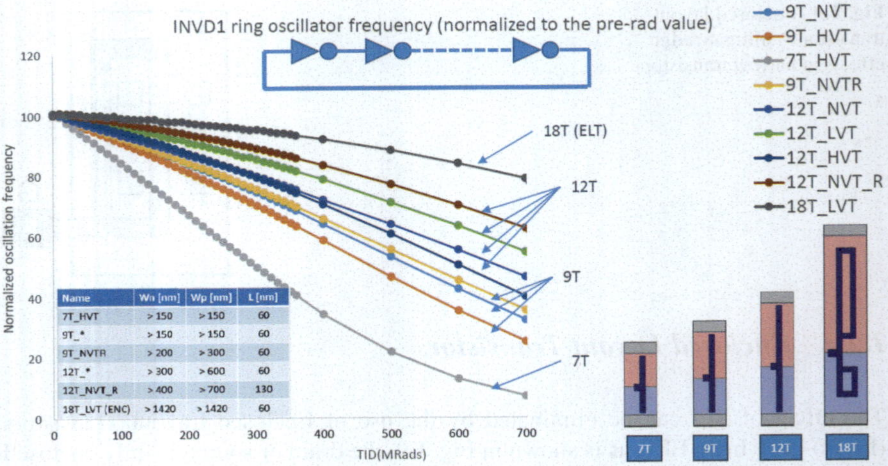

Fig. 1.10 Degradation of the speed of a standard cell library inverter in different track width for low-, normal-, and high-Vt implementations. Different widths of standard cell library transistors affect the degradation

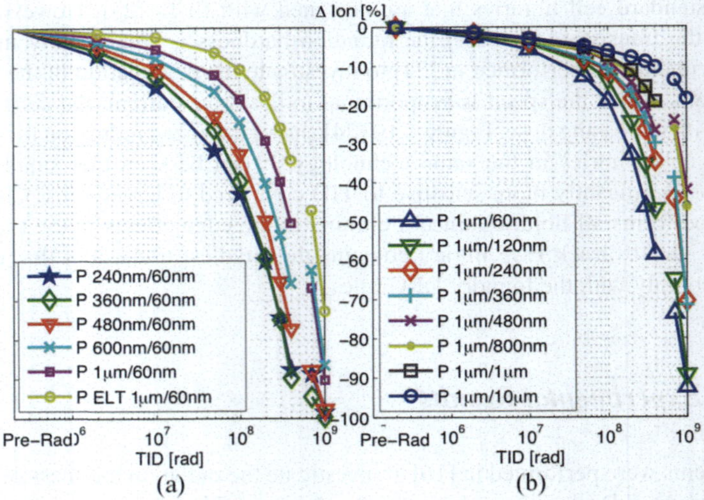

Fig. 1.11 On current of pmos transistors with different geometries: (**a**) Effect of width (**b**) Effect of length

Nmos devices show a significantly better response than pmos devices as can be seen in Fig. 1.12 [24]. For narrow channel effects, one can observe a slight increase in the current at low doses. This originates from the positive trapped charges in the STI as they improve the channel's charge. At higher doses, interface traps become

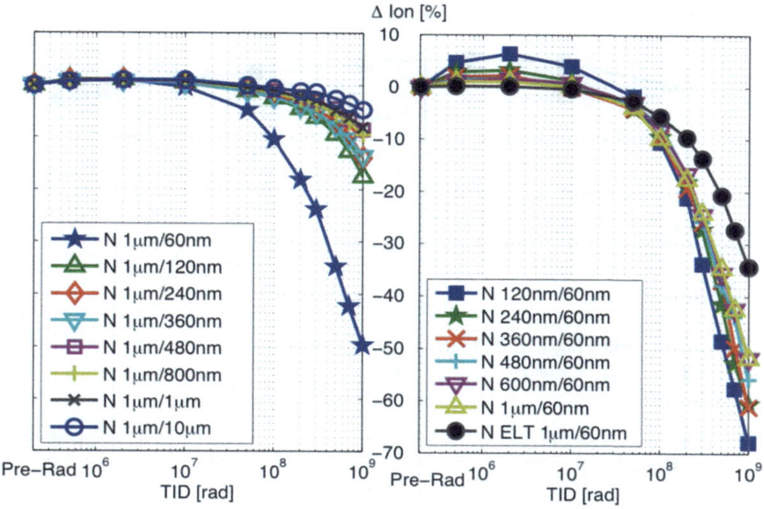

Fig. 1.12 On current of nmos transistors with different geometries: (**a**) Effect of width (**b**) Effect of length

dominant reducing the current in narrow devices. For short devices, the degradation originates from increasing hot carrier damage.

Figure 1.13 [24] shows a comparison between nmos and pmos devices with $L = 120\,\mathrm{nm}$ and $W = 1\,\mu\mathrm{m}$ which is approximately twice the minimum gate length. This demonstrates that nmos transistors are significantly more radiation tolerant than pmos devices in this technology.

1.3 Single-Event Effects

1.3.1 Basic Mechanism

Single-Event Effects (SEEs) are short-term radiation effects in the silicon. Single ionizing particles can ionize the silicon and generate electron-hole pairs as described before. While TID effects originate within the oxides, SEE mechanisms happen in the silicon [25]. The number of collected charges which were generated by the high-energy particle depends on the location of the hit. The highest collection happens when the charges are generated near the junctions of the MOS device. Figure 1.14 shows the process of an SEE hitting the silicon near the drain/source junction of a nmos transistor. After the electron-hole pairs are generated, they are separated by the high electric field in the depletion layer of the junction. When this happens, the separated charges in their turn even extend the electric field beyond the depletion

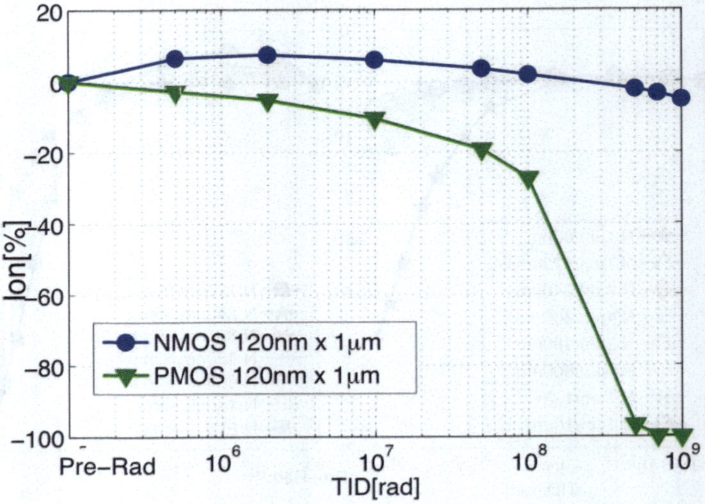

Fig. 1.13 Comparison between the current degradation in nmos and pmos transistors

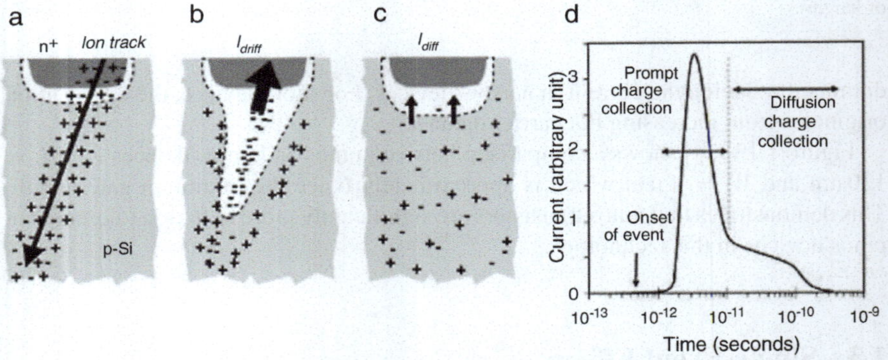

Fig. 1.14 (**a**) Ionizing particle creates electron-hole pairs in the silicon. (**b**) Electrons drift towards the n+ junction creating a funnel. (**c**) The remaining electrons and holes diffuse and recombine

layer resulting in a so-called "funnel" which can be seen as a local depletion layer. The majority of the electrons which do not recombine drift towards the n+ implant which results in a high transient current in the drain or source terminal.

When the electron-hole pairs are being separated, the funnel starts to reduce. The remaining charges which are not transported through drift will be mobilized through diffusion, which is much slower compared to the drift mechanism.

This explains the typical current waveform seen from SEEs as is shown in Fig. 1.14c. The high peak originates from drift while the longer trails originate from diffusion.

These currents are typically modeled with a double exponential Gaussian function [26]:

$$I(t) = Q\frac{e^{t/\tau_1} - e^{t/\tau_2}}{\tau_2 - \tau_1} \tag{1.4}$$

in which Q is the total charge which is collected and τ_1 and τ_2 are the time constants of the rising and falling edges. Typically, τ_1 is 10–50 ps and τ_2 rises beyond 100–200 ps. The total collected charge is difficult to predict since it highly depends on the location of the impact, doping levels, recombination rates, and the potential distribution (electric field) in the device [27].

1.3.2 Effect on nmos and pmos Devices

As discussed, in nmos devices, electrons are captured by the source/drain implants which results in a current from the source/drain junctions into the substrate/bulk contact of the device. The operation is reversed in pmos devices in which holes are captured by the p+ drain/source junctions. This leads to a current from the nwell connection to the drain/source connections as is shown in Fig. 1.15. Typically (especially in digital circuits), nwells are tied to the supply voltage while the substrate is biased at a ground potential. This results in a simplified conclusion that nmos devices drain current to the substrate from a sensitive node while pmos transistors source current to the sensitive node.

A CMOS inverter is shown as an example in Fig. 1.16. The output node is a sensitive node since it is connected to the drains of the nmos and pmos. However, when the output is a logic 0, transient currents from the nmos can never result in a 0 to 1 transition in the logic. This can only originate from a pmos device. The opposite holds true for a 1 to 0 transition.

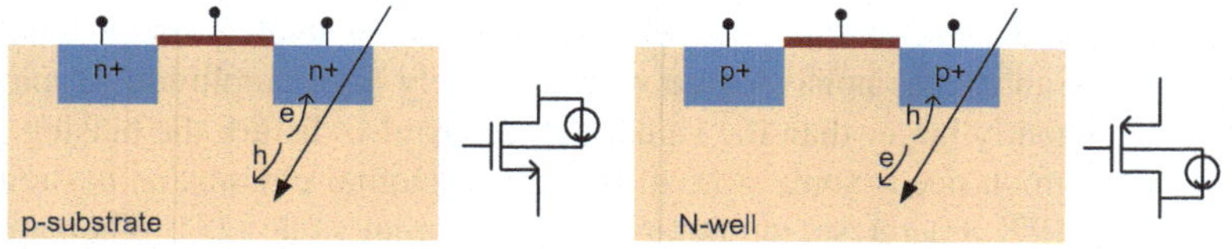

Fig. 1.15 In nmos transistors, electrons drift towards the source/drain junctions leading to a sinked current to the source/drain. In pmos transistors, holes drift towards the source/drain junction leading to a sourced current to the source/drain

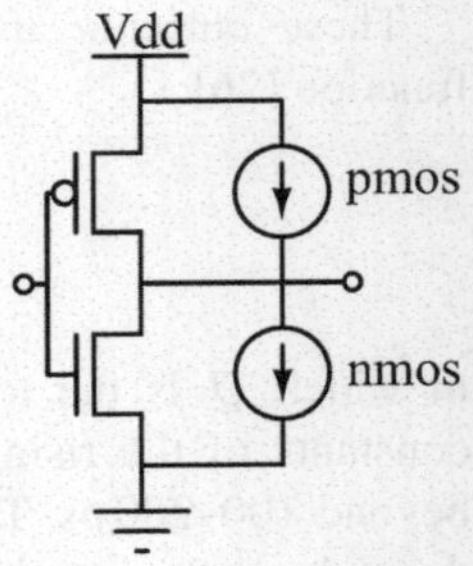

Fig. 1.16 In a digital logic CMOS cell, pmos transistors can only source current to the output node while nmos transistors can only sink current from the output node. Therefore, SEEs on nmos devices can create a 1–0 transition but not a 0–1 transition. The inverse is true for pmos devices

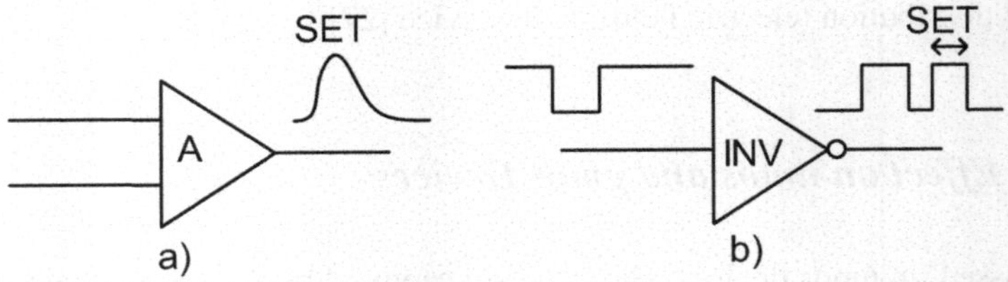

Fig. 1.17 (**a**) SETs on analog devices like amplifiers create temporary signal disturbances which lead to a dramatic reduction in SNR. In most cases, the signals cannot be used anymore. The shape of the SET may also be dependent on the impedance of the node. (**b**) In digital circuits, an SET leads to a temporary inverted bit flip in the logic

1.3.3 SET, SEU, SEL

Single-Event currents can lead to various errors in the circuits. Depending on the type of effect they have, they can be generally divided into 3 major types. Single-Event Transients (SETs), Single-Event Upsets (SEUs), and Single-Event Latchup (SEL).

SETs are transient voltages and currents which originate from the currents generated by the charged particle. In analog circuits, this leads to current and voltage transients in the circuits which may temporarily disturb the operation of the signal. Figure 1.17a shows an example of an SET occurring in an analog circuit. The SET can be seen as an excess noise source with extremely high amplitude. It may not only be significantly larger than the signal but it may also disturb the biasing point of the circuit which needs some time to recover. In digital circuits, as is shown in Fig. 1.17b, an SET generates a temporarily wrong digital value [28]. When an SET occurs in a sequential digital logic within the setup and hold times of the registers, the SET may be captured which leads to an incorrect logical state.

The sensitivity of a digital sequential block to SETs is typically dependent on three parameters. Firstly, the cross section of the logic cells which describes how much area is sensitive to the radiation. Secondly, when the clock frequency increases, the probability of capturing an SET is also increases. And thirdly, due to bit masking, not all SETs propagate to the input of a register [29]. Typically, the bit masking in digital designs approximates 40 %. The overall cross section of the design that is sensitive to SETs can be calculated as

$$\sum X_{cell} A_{mask} \frac{T_{setup+hold}}{T_{clock}} \tag{1.5}$$

in which X_{cell} is the cross section of a logic cell, A_{mask} is the digital bit masking, $T_{setup+hold}$ is the setup and hold time of the end point, and T_{clock} is the clock period.

SEUs are errors in digital circuits which have a memory-like register behavior like latches and flip-flops. When the register involves a bit-flip, this erroneous number may remain in the digital block and may even propagate to other digital modules [30]. For example, an SEU may change the state of an FSM temporarily. SEUs can originate from direct upsets in the registers but may also be a result from SETs in the combinational logic. In large on-chip SRAM memories, SEUs can severely corrupt the data in the memory. Therefore, typically scrubbing and error correction is done to refresh the memories.

CMOS technologies on silicon substrates are known to be prone to latch-up. This effect occurs due to parasitic combined bipolar transistors which originate between nmos and pmos transistors. The combined bipolar structure can be in a positive feedback state once triggered such that a DC current flows from the supply to the ground terminal. Once triggered, the only way to stop the latch-up is to cut the power supply. Generally, latch-up can be triggered through transient currents, in the PNPN structure. This is exactly what occurs when an ionizing particle passes through the chip [31]. This effect is known as Single-Event Latch-up. Latch-up can be avoided by design by ensuring small substrate resistances and wide spacing between pmos and nmos devices. However, in the technologies used in this work, SEL has not been observed for a minimum sized spacing between nmos and pmos devices in digital standard cell libraries.

These effects are all non-destructive (however, SEL can be destructive if the current density becomes too high). Single-events can also be destructive like Single-Event Burnout (SEB) and Single-Event Gate Damage (SEGD) [32] but these are not further discussed here.

Note that also non-nuclear applications cope with single-event soft errors with shrinking technology nodes [33].

1.3.4 SEU Mitigation Techniques in Digital Blocks

Digital circuits are the most easy to harden against SEEs [6]. A common and highly efficient way of ensuring correct behavior is the implementation of Triple Modular Redundancy (TMR). In a TMR circuit, all logic is implemented three times and voters are employed to correct single errors in the logic. Multiple implementations of TMR are possible depending on the available recourses [34]. The most complete form, and also the most robust is shown in Fig. 1.18. This is a general form of digital sequential logic. Both the logic and the registers are triplicated and cross-checked

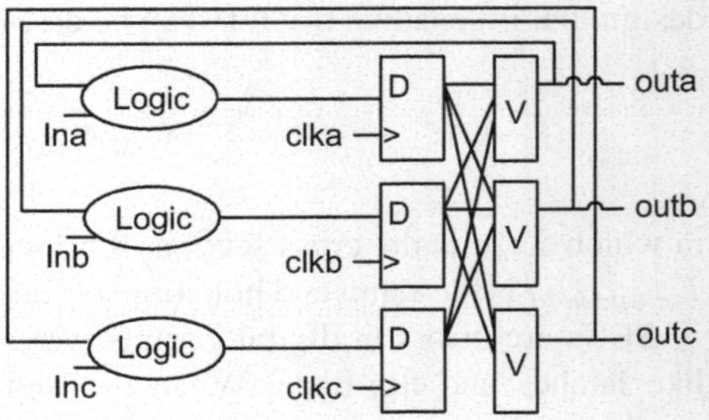

Fig. 1.18 Fully TMR structure with triplicated registers, logic, voters, and clock tree

by three voters. In case of any single error in the logic or the registers, all registers will be corrected in the next clock cycle. This technique works if there is only one error in the logic feedback path within a clock cycle. However, the probability of two upsets at the same location by different particles is extremely small. With shrinking technology nodes, the effects of multi-bit upsets become more important and may need to be addressed separately in the future. It may require different placement strategies that take into account multi-bit upsets in the logic. Note that also the clock trees for the registers are triplicated since SETs can also occur in the clock tree.

While this technique is extremely robust against SEEs, its main drawback is the power consumption and area consumption which is more than three times larger. Also, the digital timing is degraded due to the additional voters and larger routing overhead within the circuit.

Various simplified topologies, originating from this structure can be implemented which improve the area efficiency, power consumption, or speed at the cost of higher SEU cross section. For example, the triplicated clock trees can be combined in one clock tree which saves power since a lot of power is typically consumed in the clock tree. Furthermore, the combinational logic can be simplified to only one instance. As explained before, SETs in the logic depend on the propagation probability and are only captured when they happen within the setup and hold times of the flip-flop. A simplified TMR implementation may save power and area at the cost of single-event sensitivity. In this case, the design should tolerate errors at a system level since it is not fully protected.

Finally, a regular logic implementation can be used with radiation hardened flip-flops which are designed to be tolerant SEEs. For example, DICE latch based flip-flops as shown in Fig. 1.19 use interlocked nodes to correct SEUs in the latch [35]. In this circuit, nodes X1 and X3 are the inverse of nodes X2 and X4. Any upset in X1-4 will be compensated by the other nodes. Another technique which is widely used in industry is time-protected flip-flops [36]. Also, triplicated self-correcting latches are used which immediately correct themselves after an error is detected without the need to wait for a new clock cycle as is the case in regular TMR. The main disadvantage of these techniques is the need to design a custom standard cell

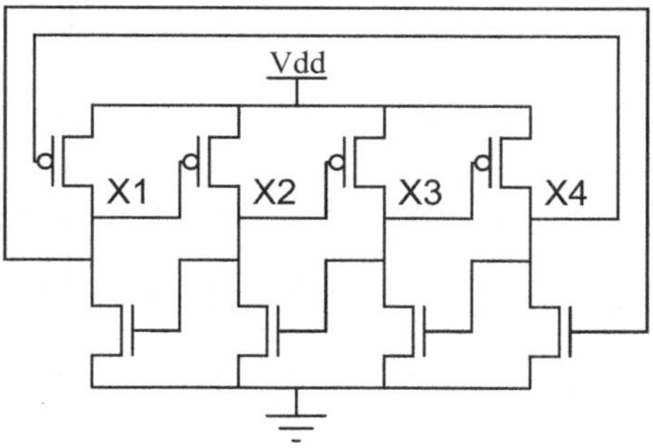

Fig. 1.19 DICE latch with four interlocked nodes

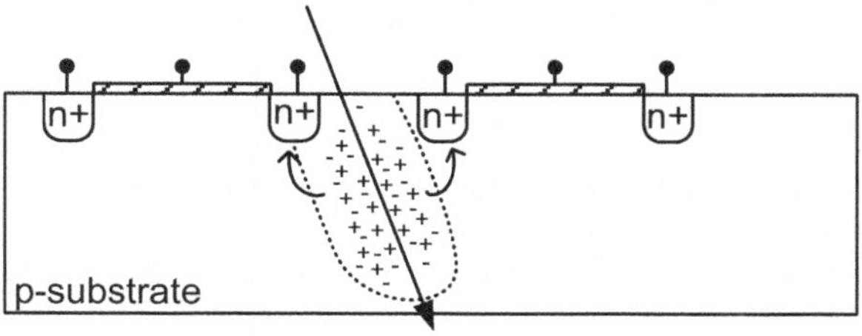

Fig. 1.20 Charge generated from a single particle can extend up to 10 μm such that it can be collected by multiple junctions. The amount of charge sharing depends on the position and reverse junction voltage of both junctions

library and extract the correct timing for digital place & route tools which may be time-consuming and expensive.

Note that TMR can be automatically generated from a HDL digital design [37].

1.3.5 Charge Sharing

The charges generated in the chip are typically generated in a diameter of 1 μm up to 10 μm perpendicular to the particles' trajectories [38]. In deep submicron technologies, this charge distribution may influence different junctions or even different transistors as is seen in Fig. 1.20 [39]. In digital circuits, multiple cells or registers may be upset simultaneously leading to Multiple-Bit Upsets (MBUs). Generally, MBUs are not disastrous when triplication techniques are used. However, TMR relies on the fact that only one bit is upset from a common register. To ensure a correct TMR operation, the three common registers (representing 1 logical bit) should be spaced enough to reduce the chances of MBUs. This can be automatically ensured in place & route tools.

In analog circuits, this effect can even be exploited to improve the radiation hardness of a circuit. For example, if the drains of a differential amplifier's input pair are closely spaced to each other, the collected charge will be shared between the nodes leading to a common mode component of the SET which will have less influence on the signal. In general, if an optimized common centroid layout is used in differential analog circuits, this effect can be exploited which is discussed in detail in [38].

1.4 Simulation Methods to Simulate Radiation Effects

Simulating radiation effects can be a challenge in complex systems. Depending on the hierarchy, system level failures can be simulated such as fault tolerance and fault propagation. At low level devices, where this work is focusing on, the goal is to find sensitive blocks in the design and the circuits. To investigate the influence of radiation on the circuit performance, simulation methods can be divided into TID effects and SEU effects.

1.4.1 Simulation of TID Effects on Circuits

TID effects are cumulative effects on the devices and degrade the devices in the circuit. As discussed before, it has become difficult to model the TID radiation effects on the nanoscale devices. A general way to model radiation effects during simulation is to extend the corners of the process in an extra dose dimension. For a given technology, the devices are characterized at different doses for different device sizes that may be available since the geometry of the devices has significant influence on the radiation sensitivity. Thus, the transistor parameters in the models are adjusted for different doses which is done experimentally. An extrapolation can be performed for different process corners in the technology.

The major challenge in this method is to correctly extract the model parameters of the devices which were measured in the experiment. Typically only device currents are measured experimentally from which other model parameters (such as channel mobility and threshold voltage) should be extracted. Figure 1.21 shows a schematic representation of the parameter extraction from a TID experiment. Before radiation, the parameters (like Vt) are known and spread over different process variations. A batch of samples, usually in a typical corner, is irradiated and measured at discrete doses. The variation can then be interpolated for different doses and extrapolated to different corners if time and budget does not allow irradiation campaigns on various process corners. Note that the dose itself does not only determine the

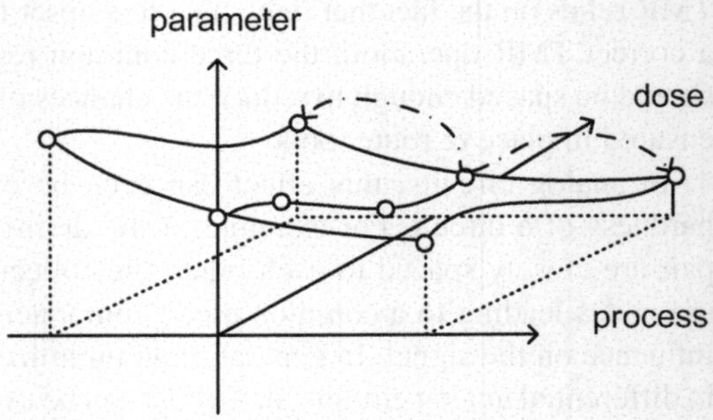

Fig. 1.21 Parameter model as function of process variations and dose

damaging effects but also the dose-rate, temperature, and node voltages have an effect on the radiation damage. These parameters will be additional dimensions in this model.

1.4.2 Simulation of Single-Event Effects on Circuits

SEU effects are intrinsically different from TID effects since the timescale is different and the effect is not accumulating. Once the correct parameters are extracted from TID experiments, the simulation method is straightforward as an extra corner dimension. TID effects can be assumed to be uniform for the different devices on the same chip which is not the case for single-event effects as they happen for only some fraction of time locally.

In digital circuits, SEU effects can be estimated by fault injection methods. In digital simulators, random bit-flips and transients can be inserted to investigate if the system is sensitive to the upsets.

In analog blocks, the amount of charge captured by the devices can be important since it disturbs analog valued signals, also the shape of the SEE current should not be neglected. These shapes can be estimated from TCAD simulations which can calculate charge mobility in the devices after a particle has hit the chip [40]. The amount of charges and the interaction of the particle with the silicon can be estimated from Monte-Carlo nuclear simulators like GEANT4 [41] and FLUKA. These are tools which are developed at CERN to simulate the interaction between different radiation sources and material geometries. The output of these tools are 3D (or 2.5D) energy depositions of the particle in the silicon from which the amount of electron-hole pairs being generated can be calculated. This data can be used in TCAD to simulate voltages and currents in basic circuits with few transistors. To simulate the analog blocks in a larger system, the transients should be modeled to behavioral models of the components. Figure 1.22 shows a schematic on the

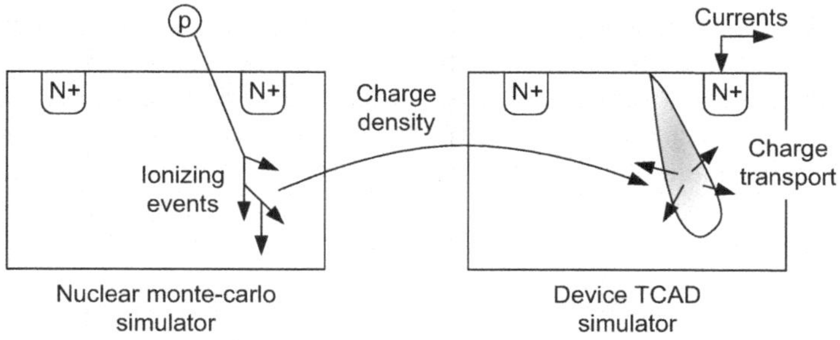

Fig. 1.22 Monte-Carlo particle interactions estimate energy deposition in ionization events in the silicon substrate from which charge generation can be estimated. The charge density distribution is the initial state of carrier generation for the TCAD model

SEU simulation for spice input. In the first step, MC nuclear simulator estimates the spatial energy deposition and charge densities which are initial states for the TCAD simulator. The output of this flow are currents at the device nodes which can be used in spice simulations [42]. Note that this simulation way is very time-consuming and is an estimation for the reaction of the circuit to the charges since correct TCAD models cannot be constructed since the foundry information is not made available. Therefore, this flow is used to understand the mechanisms how the circuits react to the charges and not as a formal verification of the device performance. Radiation assessment is still required for the devices before they can be qualified for applications in radiation environments.

These methods can be time-consuming and impractical in many designs. As discussed before, the current pulses can also be modeled as a double exponential function [26].

1.5 Conclusion

Long term, accumulating radiation effects are observed, even in advanced CMOS technology nodes. While they were originally observed in the gate-oxide, TID effects today are seen in secondary oxides such as STI and LDD spacer oxides. In 65 nm CMOS technology nodes, it has been seen that pmos transistors degrade significantly more than nmos devices. Also, TID effects were observed to be significantly geometry dependent such that short and narrow transistors degrade more than long and wide transistors.

Single-Event Effects become even more important in shrinking technologies as the node capacitances are decreasing. Ionizing particles generate temporary disturbing charges and currents in the circuits which may lead to transient voltages and incorrect digital values. The charges are collected at the device junctions. While charge sharing can occur between different nodes, triplicated structures are a common approach to mitigate SEEs in digital modules at the cost of power and area.

While radiation cannot be directly included in SPICE simulations, an approximation through Monte-Carlo simulations and TCAD models can be made to estimate the currents generated in the devices to simulate the effect on the circuits. TID effects were known to be successfully included as additional dimension in the corners of the technology.

Chapter 2
Time-Domain Signal Processing

2.1 Introduction

Integrated circuits and CMOS technologies are mainly driven by technology scaling. In most systems, digital circuits occupy the largest portion in terms of power, area, and complexity. Digital circuits take full advantage of shrinking of devices because the power consumption reduces and the functionality (larger number of devices on a single chip) and switching speed increase [43]. However, analog circuits suffer from reduced intrinsic gains and reducing supply voltages, especially below 90 nm nodes where the supply voltage approaches 1 V. This has led to highly digital architectures which are faster, occupy less area, and are less power hungry. Obviously, some functions can be and still need to be done with analog circuits but digital circuits are employed wherever possible [44, 45].

Information can be contained in many physical forms, depending on the origin. In analog electrical circuitry, most of the information is represented with currents and voltages. While this may be a straightforward way, it has been proven that time can be used as well to maintain signal information. As technology nodes are shrinking, so does the supply voltages of the devices which makes it harder to properly design analog, voltage processing circuits with high swings and good power supply rejection and linearity. Time-domain processing involves Time-to-Digital converters (TDCs), which are accurate stopwatches which may benefit from faster digital blocks since the resolution of a TDC has the potential to improve with each technology step [46]. TDCs convert time differences to a quantized digital word which can be further processed by a digital core. The TDC is comparable to an ADC acting in the voltage domain. Most TDC architectures are implemented using delay lines which explains why these highly digital architectures benefit from CMOS scaling.

In general, the traditional interface to the analog world is shown in Fig. 2.1. The interface to the analog world typically involves a critical pre-processing circuit

J. Prinzie et al., *Radiation Hardened CMOS Integrated Circuits for Time-Based Signal Processing*, Analog Circuits and Signal Processing,
https://doi.org/10.1007/978-3-319-78616-2_2

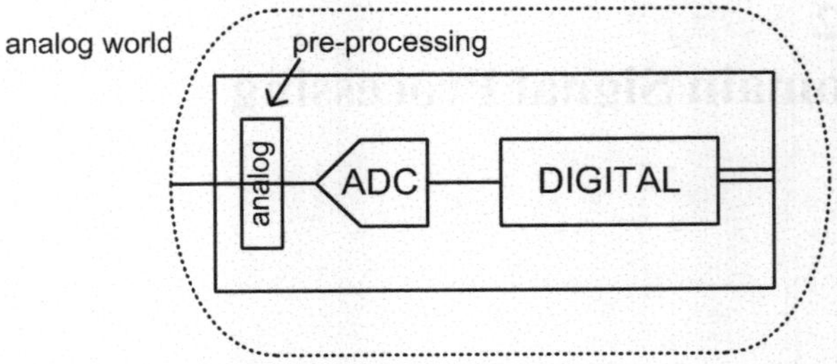

Fig. 2.1 Generic analog voltage interface

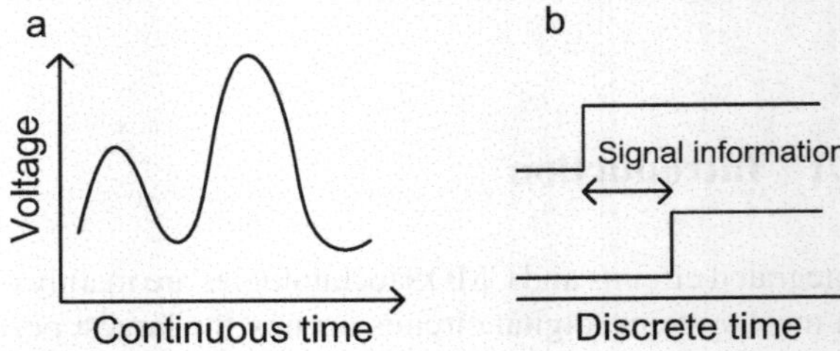

Fig. 2.2 (**a**) Voltage signal processing. (**b**) Time-domain signal processing

which requires low noise and high dynamic ranges. For example, a low noise or variable gain amplifier precedes the ADC to match the latter to the signal range specifications of the system. When time-signal processing is employed, where the information is maintained in the timing of the input signals, the ADC is implemented as a TDC and this pre-processing is typically not required.

A comparison between voltage-signal processing and time-signal processing is shown in Fig. 2.2. In the voltage domain, information can be represented continuously, both in voltage and time. After the ADC, the information is quantized both in time and amplitude. However, in time-domain processing, as shown in Fig. 2.2b, the information can only be represented in discrete time. Time domain signals are represented by time differences between two signals or two events. As these events cannot occur at infinite rate, the information stream will always be quantized in time. However, the amplitude of the information (by means of time intervals) can still be continuous as the time difference between two events is a continuous variable before the TDC. The output of the TDC is discrete both in time and amplitude.

2.2 Time-to-Digital Converters

TDCs are circuits acting as high-precision stopwatches which can measure time differences with a resolution down to a few picoseconds [47]. The input signals to the TDC act as start and stop events to the stopwatch and the output of the TDC is

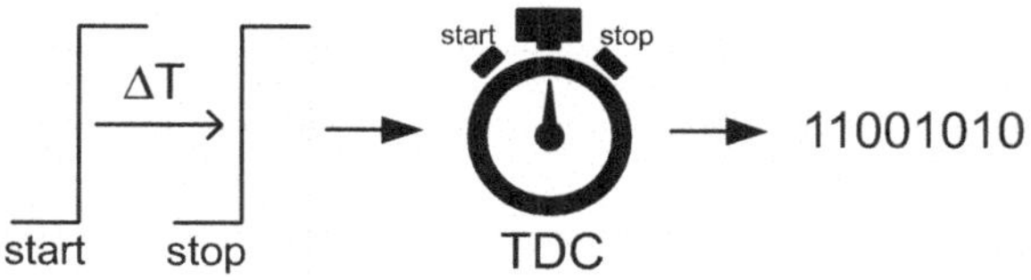

Fig. 2.3 Concept of a TDC digitizing a time difference between two digital signals

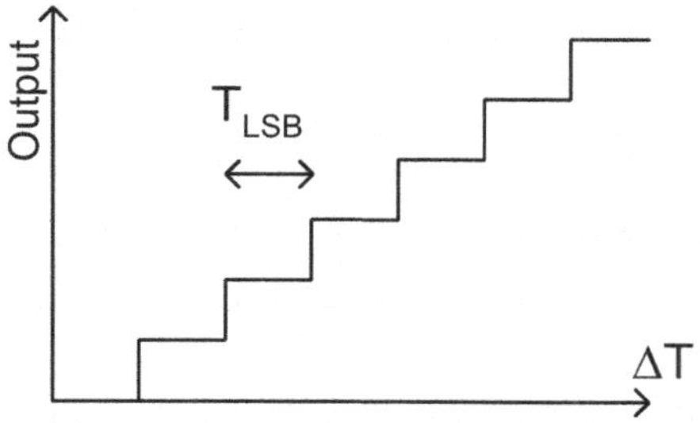

Fig. 2.4 Ideal transfer function of a TDC. The width of one bin is T_{LSB} and represents the finest resolution that may be read from the TDC

a digital word representing the time difference between start and stop as illustrated in Fig. 2.3. Since the output of the converter is a digital word, it can only represent discrete quantities of time. In other words, the output of the TDC is a quantized version of the input time difference. The analogy can be made to voltage ADCs. In an ADC, the full scale input voltage range is represented by N bits, so there are 2^N discrete intervals within this range that are represented by the binary word. A voltage step of 1 LSB is the finest resolution that may be read from the converter which is $V_{fullscale}/2^N$. In a TDC, the LSB also represents the smallest time difference that can be resolved in the TDC, ideally, by increasing the input time difference by T_{LSB}, the output word adds up by one. This leads to a similar transfer characteristic compared to an ADC as is shown in Fig. 2.4. T_{LSB} is also called the time-bin of the TDC.

It is clear that this is only an ideal transfer function and that, in reality, non-ideal effects occur. The focus of today's research is driven to improve the resolution of the TDC such that time differences can be accurately quantized to finer timing bins. Thus reducing T_{LSB}. This has been widely driven by technology improvements (improvement of logic gates) and improved architectures to downscale T_{LSB}. Furthermore, much effort is also put into non-linearities and self-calibration structures to make TDCs industrially valuable and ensure reproducibility.

The most basic TDC (Also called a "flash TDC") is shown in Fig. 2.5 and is based on a delay line [48]. The start signal is injected in the delay line and propagates into the line until the stop signal arrives. This stop signal takes a snapshot of the delay line. By measuring how far the pulse has propagated into the delay line (how many gates), the time difference is digitized. The more time has expired between start and stop, the further the start edge will propagate in the delay line.

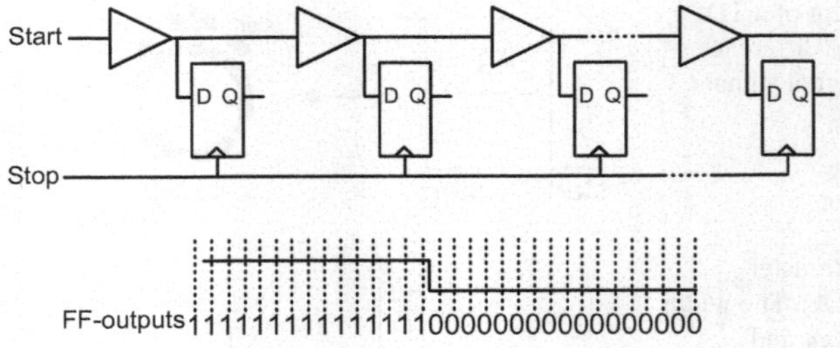

Fig. 2.5 Basic flash TDC based on a delay line which is sampled by a layer of flip-flops

2.3 Applications of Time-Based Circuits

Many TDCs have been developed in the past 20 years, mainly in the field of high-energy physics experiments in which TDCs measure time differences in particle detectors. However, with advanced technology scaling, TDCs have become interesting in other fields such as optical Time-of-Flight measurements, sensor interfaces, and All Digital PLLs.

2.3.1 High-Energy Physics

High-energy physics experiments such as ATLAS or CMS at CERN try to take a snapshot of the particles generated by proton collisions. The function of the detectors is to register the time, position, energy, and momentum of the particles as they cross the detectors. An example of one of many detectors is the CMS muon drift tube (DT) which detects the muon position in different layers around the detector. A cross section of a DT is shown in Fig. 2.6 [49]. The detector is assembled as an array of positively charged wires in a chamber which is filled with Ar and CO_2. When a muon travels through the gas, it ionizes the gas. Free electrons drift towards the positively charged wire. In the case of the CMS DT, the wires have a potential of 3600 V leading to a drift speed of 54 μm/ns. Each drift tube has a dimension of approximately 42 mm × 13 mm. The horizontal position where the particle crossed the tube is measured by the time it takes for the charges to drift towards the wire. A TDC is used to measure this time in the CMS detector. Figure 2.7 [50] shows a detailed drawing of the drift tubes. Vertical field strips are placed to improve the lateral electric field in the detector. A Monte-Carlo simulation and measurement is also shown [50]. Based on the TDC output, a position is estimated by the system.

TDCs are also used in silicon pixel chips, which are reverse biased diodes that capture charges of particles that crossed the detector [51]. A pixel chip is an array of small silicon pixels. When a pixel is hit by a high-energy particle, charges are generated in the junction (Fig. 2.8). Each pixel has a front-end amplifier and pulse-

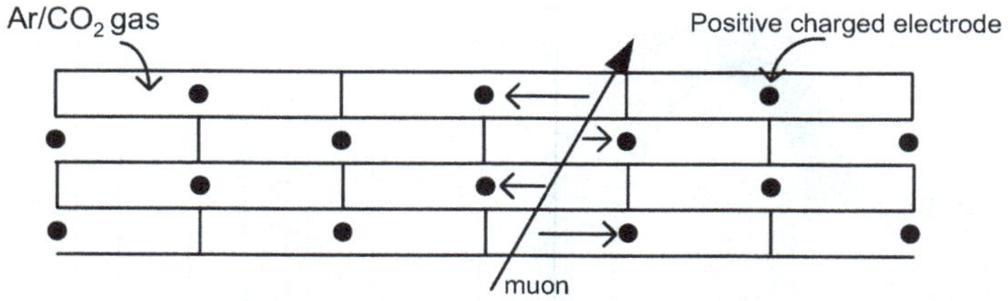

Fig. 2.6 Cross section of a drift tube detector used in the CMS experiment to measure the position of muons. Muons ionize the gas mixture and free charges drift towards the positively charged wire

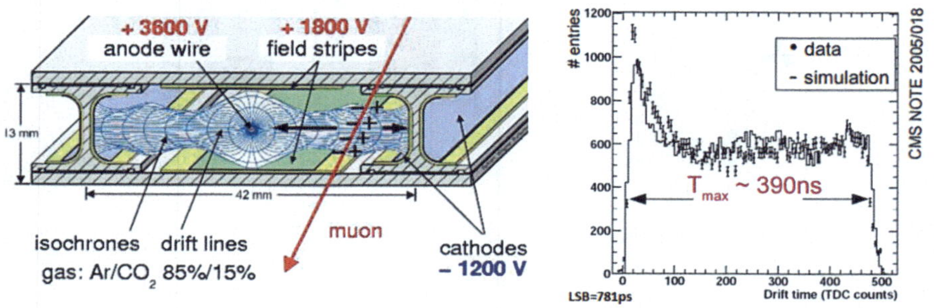

Fig. 2.7 The electric field in the detector ensures that the generated charges drift towards the wires. A TDC measures the drift time until the charges are collected by the wires. A histogram shows the output of a TDC channel for uniform particle hits in the drift tube

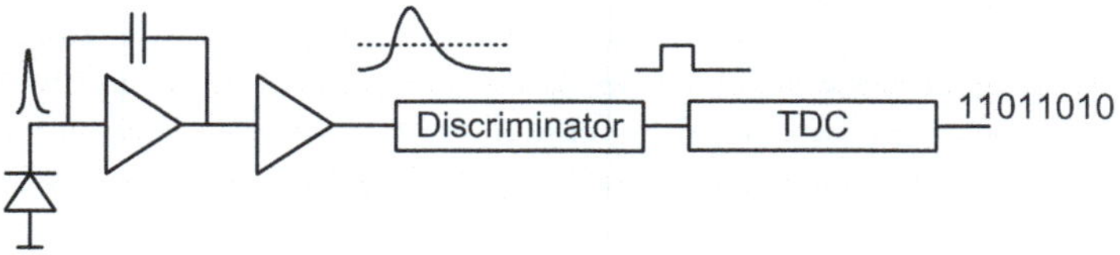

Fig. 2.8 Schematics of a pixel front-end. A charge integrator collects the charge from the silicon pixel. A shaper generates a pulse based on the charge integrator. A discriminator generates a pulse to measure the time-over-threshold

shaper. These analog front-end electronics generate an output pulse based on the charge collected by the pixel. The amount of charge that is collected is measured by studying the shape of the pulse, generated by the front-end amplifier and shaper. Most pixel chips measure the time the pulse is above a given threshold, this is also called "Time-over-Threshold" (TOT) measurements. The more charge is injected to the front-end, the higher the pulse will be and the longer the signal will be above the threshold. TDCs are used to measure this TOT and are much more power and data efficient than processing the signal in the voltage domain [52].

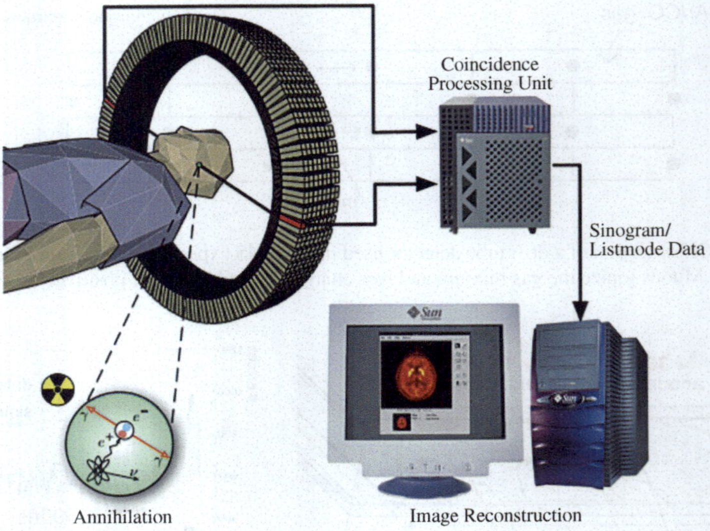

Fig. 2.9 PET scanner illustration

2.3.2 PET Scanners

Positron Emission Tomography (PET) scanners are medical imaging systems that work, based on radioactive isotopes which are injected in a patient [53]. The fluid is absorbed by the blood system and in a later stage, absorbed by the organs. The isotopes decay, generating positrons inside the patient as is shown in Fig. 2.9 [54]. When this happens, a positron will quickly annihilate with an electron generating two photons which have opposite momentum, this means one photon travels in one direction while the other travels in the other way. Around the patient, a circular array of detectors, sensitive to these gamma photons is placed. The actual position from where the photons were generated can be calculated if the arrival time difference between the two photons is known. TDCs are used to measure this time difference. The photon detectors are based on scintillator crystals [55] which convert the photon energy to visible light that can be detected by a photo-diode.

2.3.3 Time-of-Flight LIDAR

Distance measurement using light is done through LIDAR (Light detection and ranging). The technique is comparable with RADAR but uses light instead of radio waves. When light is emitted by a laser, it travels through space at the speed of light. When an object interferes with this light, part of it will be reflected back. Time-of-

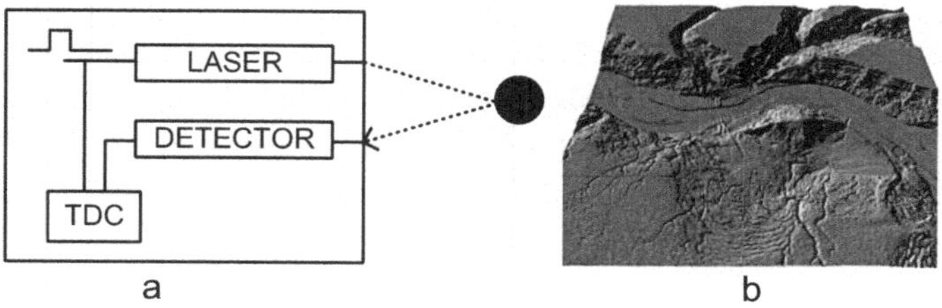

Fig. 2.10 (**a**) LIDAR system architecture. (**b**) Example output of a LIDAR

Flight LIDAR works by emitting pulses of light and observing the time it takes for the reflection to arrive [56]. From this arrival time, the distance can be calculated with

$$d = \frac{c.t}{2} \tag{2.1}$$

in which d is the distance, t is the time of flight, and c is the speed of light (in the propagation medium).

A simplified architecture of a LIDAR is shown in Fig. 2.10a where a TDC measures the time difference between the transmitted time and the arrival time of the light. In reality, LIDAR systems are much more complex to deal with the delays in the laser and photo-receivers and include photomultipliers to measure very weak reflections [57].

If a measurement resolution of 1 mm is required, the light will travel approximately 6 ps (forward and backward) in this interval in free space. It is clear that very accurate timing resolutions are required. However, these experiments can be repeated many times if the objects are fixed. Oversampling techniques can be used to relax the constraints on the system. For example, sigma-delta based TDCs can be used as discussed in Sect. 2.4.

These circuits have been proven radiation hard up to 5 MGy [58].

2.3.4 All-Digital PLLs

Originally, Phase Locked Loops (PLLs) operate in an analog- mixed signal way with a digital phase-frequency detector (PFD) and an analog charge-pump and loop filter which control a voltage controlled oscillator. However, it has become interesting to also implement the charge pump and loop filter in the digital domain to eliminate process variations in the analog blocks and noise generated in the passive and active devices. In this way, the analog filter is replaced with a digital filter which controls a digitally controlled oscillator (DCO) instead of a voltage controlled oscillator. Originally, the PFD, in combination with the charge-pump, generated correction signals proportional to the phase difference between the output phase and the reference phase. In all-digital PLLs (ADPLLs), the phase measurement is done by a TDC and the digital output value is used by the digital loop filter to correct the oscillator. This approach has been widely investigated and shows good performance in frequency synthesizers [59].

2.4 TDC Circuits

2.4.1 *Performance Parameters*

2.4.1.1 Single-Shot Precision

Single-shot precision is the accuracy that a TDC can achieve if the measurement is done only once and when noise is considered in the TDC. To find this precision, the probability that the TDC makes an error has to be calculated so the error can be weighted by its probability [47].

Consider a noisy timing inaccuracy (jitter) which adds to the measurement time-interval T_{in}. The probability density function of this jitter is modeled by a Gaussian distribution:

$$p_{\tau_j}(\tau_j) = \frac{1}{\sqrt{2\pi}\sigma_{\tau_j}} exp\left(-\frac{\tau_j^2}{2\sigma_{\tau_j}^2}\right) \tag{2.2}$$

If an input time difference T_{in} is considered within the range of $\pm\frac{1}{2}T_{LSB}$ (Fig. 2.11), the probability can be calculated for which the TDC will not decide that this measurement is in the current bin but in the next, previous or even further bins. For this range, the decided value should be zero but due to noise, this may not be always the case.

A function $p_n(T_{in})$ can be defined which describes the probability that the TDC makes an incorrect decision to bin n. If the jitter τ_j is within the following intervals, the incorrect bin decided by the TDC is as follows:

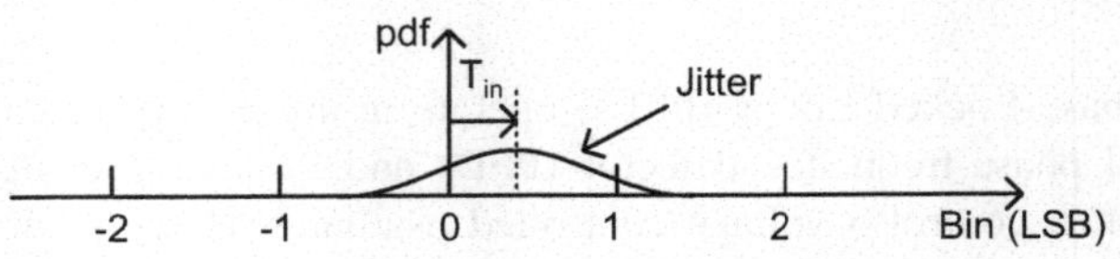

Fig. 2.11 Graphical representation of an input time difference with jitter in the TDC

$$
\begin{aligned}
-\frac{3}{2}T_{LSB} \leq T_{in} + \tau_j < -\frac{1}{2}T_{LSB} \quad &\rightarrow \quad n = -1 \\
-\frac{1}{2}T_{LSB} \leq T_{in} + \tau_j < \frac{1}{2}T_{LSB} \quad &\rightarrow \quad n = 0 \\
\frac{1}{2}T_{LSB} \leq T_{in} + \tau_j < \frac{3}{2}T_{LSB} \quad &\rightarrow \quad n = +1
\end{aligned} \tag{2.3}
$$

So, the TDC will decide bin n if

$$
\tau_j \in \left[\left(n - \frac{1}{2}\right) T_{LSB} - T_{in}, \left(n + \frac{1}{2}\right) T_{LSB} - T_{in}\right] \tag{2.4}
$$

Note that the actual value should be bin number zero.
$p_n(T_{in})$ can then be formulated as

$$
\begin{aligned}
p_n(T_{in}) &= \int_{(n-\frac{1}{2})T_{LSB}-T_{in}}^{(n+\frac{1}{2})T_{LSB}-T_{in}} \frac{1}{\sqrt{2\pi}\sigma_{\tau_j}} exp\left(-\frac{\tau_j^2}{2\sigma_{\tau_j}^2}\right) d\tau_j \\
&= \frac{1}{2} erf\left(\frac{(n+\frac{1}{2}) - T_{in}}{\sqrt{2}\sigma_{\tau_j}}\right) - \frac{1}{2} erf\left(\frac{(n-\frac{1}{2}) - T_{in}}{\sqrt{2}\sigma_{\tau_j}}\right)
\end{aligned} \tag{2.5}
$$

If the jitter σ_{τ_j} and the input time difference are given, the probability for each bin can be calculated with $p_n(T_{in})$.

Therefore, the average expected quantized output value $\langle T_q \rangle$ of the TDC for an input time difference T_{in} equals

$$
\langle T_q(T_{in}) \rangle = T_{LSB} \sum_{n=-\infty}^{+\infty} n.p_n(T_{in}) \tag{2.6}
$$

Its standard deviation or the single-shot precision of the TDC is then

$$
T_{SSP}(T_{in}) = \sqrt{\sum_{n=-\infty}^{+\infty} (n \cdot T_{LSB} - \langle T_q(T_{in}) \rangle)^2 . p_n(T_{in})} \tag{2.7}
$$

The probability density function $p_n(T_{in})$ is shown in Fig. 2.12a for $n = 0$. The plot shows the PDF for different values of jitter. For low jitter values, the PDF approaches a rectangle. The probability that a wrong decision is made if the input time difference is T_{in} for bin n can be found by taking the value of the PDF at $T_{in} - nT_{LSB}$. When the jitter is small, the PDF is near zero for $n \neq 0$ but becomes nonzero when the jitter increases.

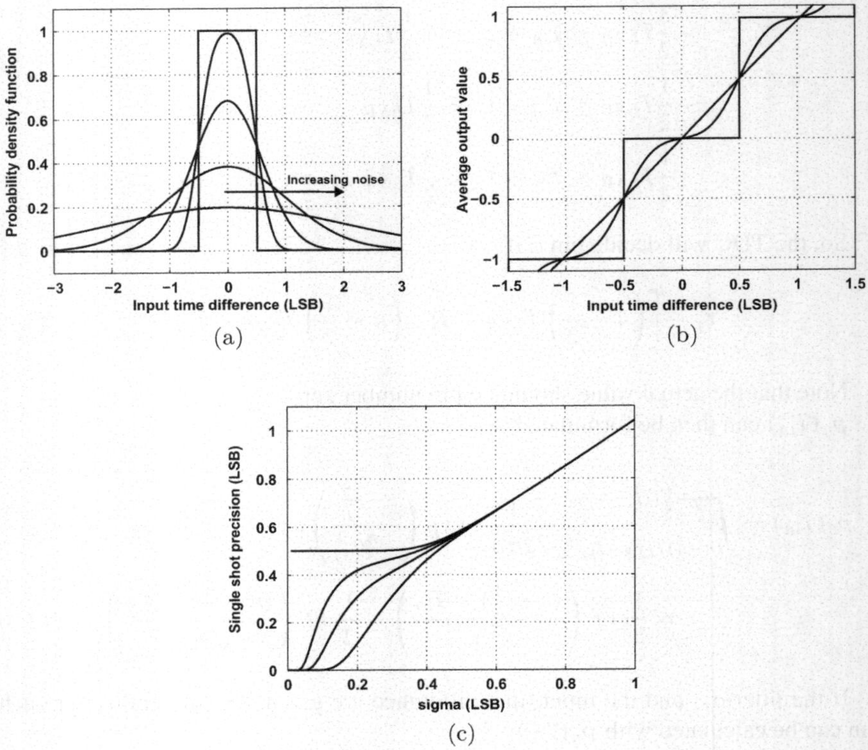

Fig. 2.12 Quantization effects with noise. (**a**) $p_n(T)$ for $n = 0$. (**b**) Average output value of the TDC for different values of jitter. (**c**) Single-shot precision for different input time differences

Figure 2.12b shows the transfer function of the average output number for a given input time difference according to (2.6). When no jitter is applied, the transfer function is step-wise but is smoothened when jitter is applied. With excess amount of jitter, the transfer function becomes a straight line.

The single-shot precision according to (2.7) is shown in Fig. 2.12c which shows the single-shot precision for different input time differences from 0 up to 0.5 LSB. If the T_{in} is 0.5, the single-shot precision approaches 0.5 when the jitter is small while the precision goes to zero for smaller input time differences. At large amounts of jitter, all curves converge as the error due to erroneous TDC decisions becomes larger than the quantization. This can be explained as follows. If the time difference is close to the center of the interval, for small amounts of jitter, the probability is small that the decision will end in the neighboring bins. However, if the input is at the edge of the bin, there is 0.5 probability that the value is wrong.

2.4.1.2 Linearity

Every TDC will have internal imperfections between the different elements. Referring to the flash TDC, it is clear that not all delay cells have an equal propagation delay and mismatch occurs between the timing of the different cells. This will contribute to non-linearities in the transfer function of the TDC. Besides mismatch in the cells, also mismatch between the flip-flops and its clock tree will contribute to errors.

In a data converter, two types of errors are typically defined: DNL and INL errors. DNL (differential non-linearity) is the difference between the actual bin width and the ideal bin width. Ideally, each bin is 1 LSB wide. However, some bins will be smaller, some will be wider.

INL (integral non-linearity) errors are the differences between the actual crossing points of the transfer function and the ideal ones.

If a simple delay line is considered, each delay element can be modeled as a nominal delay with a random offset [60].

$$T_i = T_{LSB} + \epsilon_i \tag{2.8}$$

in which ϵ_i is a random, constant delay for each cell. The delay of a delay line after n stages is the sum of these individual delays.

$$T_N = \sum_{i=1}^{N} T_i = N \cdot T_{LSB} + \sum_{i=1}^{N} \epsilon_i \tag{2.9}$$

The standard deviation of the delay after n stages, considering no correlation is then

$$\sigma_{T_n} = \sqrt{n}\sigma_{\epsilon_i} \tag{2.10}$$

This means that the timing uncertainty in a delay line increases along the line. The DNL error of a single delay line is simply

$$DNL_i = \frac{T_{i+1} - T_i - T_{LSB}}{T_{LSB}} = \frac{\epsilon_i}{T_{LSB}} \tag{2.11}$$

From (2.10), it is clear that the uncertainty or INL error accumulates in the delay line. Longer delay lines suffer more from INL errors than short delay lines.

2.4.1.3 Gain Error

The gain of a TDC can be defined by linearizing the slope of the transfer function. Since most TDCs are based on delay lines, the gain of the TDC is highly dependent on PVT variations. Slow corners provide low gain while fast corners provide high gain if no calibration is used. These effects can be compensated by a reference calibration or with a Delay Locked Loop.

2.4.1.4 Conversion Speed

TDCs can only convert time at a finite rate. The conversion of a measurement requires some time. Firstly, the time between start and stop is required which is a fundamental limit. This means that the conversion speed is limited by the dynamic range. Secondly, after the measurement is done, the data may need to be decoded. Thirdly, the delay lines may require new initialization that results in a latency.

2.4.2 *Delay-Line Based TDCs*

Flash TDCs, as discussed earlier, are based on a delay line which delays the start signal. The finest time differences that can be measured are limited by the delays of the cells. In this architecture $T_{LSB} = T_{delay}$. The dynamic range of a flash TDC is simply determined by the total delay of the delay line: $DR = NT_{LSB}$.

In Sect. 2.4.1.2 it was shown that the uncertainty in a delay line grows in progressive stages (2.10). This will limit the resolution in the TDC due to a problem with linearity at the end of the line. The second constraint is noise. Each transition in the delay line will introduce some jitter. Since the stages are cascaded, the jitter of each cell accumulates through the delay line.

If the jitter of each cell is represented by τ_{j_i}, the jitter at the end of the line will be

$$\tau_{j1:N} = \sum_{i=1}^{N} \tau_{ji} \tag{2.12}$$

If the noise in each cell is uncorrelated, the total jitter that may be expected at the output will be

$$\sigma_{\tau_{j1:N}} = \sqrt{N}\sigma_{\tau_{j_i}} \tag{2.13}$$

Note that this is only valid for random, uncorrelated noise sources and may be worse for supply noise which is correlated [61].

The advantage of this type of TDC is its simplicity but it has some drawbacks. Firstly, the dynamic range is limited by the amount of cells. Increasing the delay line size implies a reduction in linearity (higher INL) and jitter. Secondly, noninverting delay stages are proposed in this architecture which require two inverters and double the minimum LSB bin size.

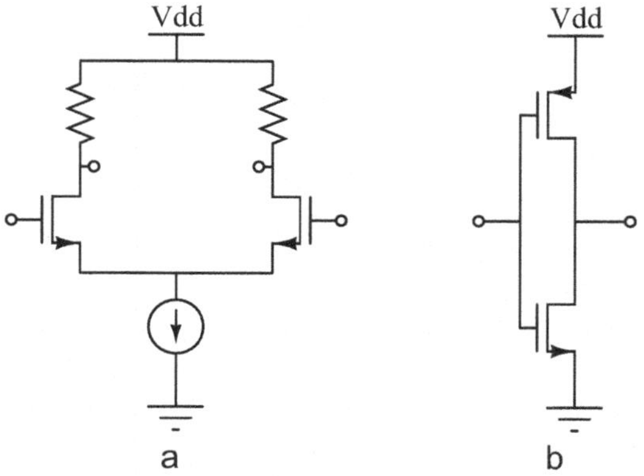

Fig. 2.13 (**a**) Differential delay cell with constant bias current. (**b**) Single-ended inverter

To overcome this, inverters can be used and the inversion can be extracted digitally. However, the TDC now relies on rising and falling edges which have to be captured by the flip-flops. This can, depending on the flip-flops' architecture result in an increase of DNL error since rising edges may have different setup and hold times compared to falling edges in some corners.

A differential or pseudo-differential architecture does simplify these inversions as it can be done by inverting a differential wire. In fully differential TDCs, both the delay line and flip-flops are differential. Differential flip-flops are usually implemented using sense amplifiers [62] as they do not consume static power. A common implementation is to use traditional CMOS flip-flops and add a differential to single-ended converter array between the delay line and the flip-flops. Finally, the delay line can also be implemented as a pseudo-differential delay line by using two delay lines with opposite polarity. This relaxes the issues of inversion as is the case with fully differential cells but may be more prone to supply noise.

If a differential cell (Fig. 2.13a) is compared to a single ended CMOS inverter (Fig. 2.13b), the jitter may become a concern. The rms jitter of a delay cell can be represented as [63]:

$$\sigma_{\tau_j} = K\sqrt{T_{delay}} \tag{2.14}$$

in which K is a constant for the circuit architecture and T_{delay} is the delay of the cell. The number K allows different circuits to be compared as the jitter depends on the total delay of a cell.

As is discussed in [63], K can be calculated for both circuits. For differential delay cells,

$$K \approx \sqrt{\frac{kT}{I_{tail}V_{swing}}} \tag{2.15}$$

in which I_{tail} is the tail current of the differential cell and V_{swing} the output swing.

For single-ended circuits, this becomes

$$K \approx \sqrt{\frac{kT}{I_{peak}(V_{DD} - V_{th})}} \tag{2.16}$$

in which I_{peak} is the current during switching and V_{th} the threshold voltage of the switching transistor.

From these findings, it can be concluded that single-ended topologies can have a better noise performance compared to differential circuits given the same power consumption as the current during switching is important. In differential stages, this current is fixed while for single-ended circuits, this current is highly peaked. However, for supply noise, differential circuits are preferred.

2.4.3 *Sub-gate Delay-Line TDCs*

The resolution of either differential or single-ended delay lines will always be limited by the minimum delay of the delay cells. To overcome this, different architectures were invented (which are based on delay lines) that allow sub-gate delay resolution.

2.4.3.1 Vernier Architecture

A commonly used circuit is a vernier delay line as is shown in Fig. 2.14 [64]. This architecture uses two delay lines from which the delays are different. The delay of the stop line is slightly faster than the delay in the start line. The point where the stop pulse catches up with the start pulse is used to measure the time difference. The readout resolution that can be achieved with a vernier TDC is $T_1 - T_2$ so the time difference determines the resolution [65]. For example, if two delay lines

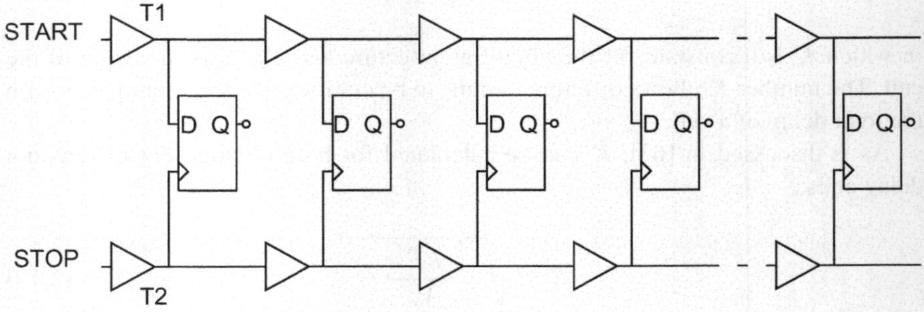

Fig. 2.14 Vernier delay line topology

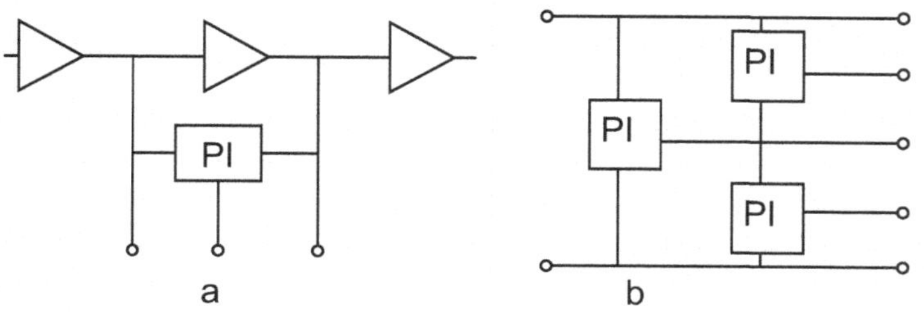

Fig. 2.15 (**a**) Phase interpolation between two delay taps. (**b**) Multi-layer phase interpolation

are used with delay elements of 15 and 20 ps, the readout resolution is 5 ps. The limitation of this architecture lies in the mismatch and jitter. Compared to a simple flash TDC, for the same dynamic range, longer delay lines are required in a vernier architecture which contribute to jitter and mismatch. This method can also be used in two dimensions [66] to extend the dynamic range of the TDC.

2.4.3.2 Local Phase Interpolation

The delays between two successive delay cells can be interpolated using a phase interpolator (Fig. 2.15a) [67, 68]. The interpolation relies on the finite slew rates of two delayed signals. The average of both signals is then time-interpolated. If the delays are interpolated once, the resolution of the TDC is doubled. This method can be applied again to the interpolated delays to further scale the delays as is shown in Fig. 2.15b. The challenges in these circuits lie in the matching between the interpolated timing bins. The implementation of the phase interpolators can be done in different ways. First, two inverters, driven by two delay taps can be connected with their outputs. This circuit will produce a waveform which has an average delay between both taps. Second, passive networks can be used which will be discussed in detail in Chap. 4. This technique relies on the relatively slow slew rise and fall times of the voltage signals. Again, the challenge lies in the matching between different interpolated delays.

2.4.3.3 Parallel TDCs

Parallel TDCs use an array of delays instead of a cascade of delay cells. Each successive branch will add a small additional delay which can be smaller than the gate delay [69]. This method can be combined with a delay line to locally interpolate gate delays.

2.4.4 Delay-Locked Loops

In general, TDCs are based on time intervals or delays from CMOS circuits. These however do vary with PVT variations or ionizing radiation effects. The variations lead to an unknown conversion gain and an undetermined readout resolution. Delay lines can be made adjustable as voltage-controlled delay lines. As will be discussed in detail in Chap. 4, a Delay-Locked Loop will adjust the delay of the line such that it is matched to the period of a reference clock. In this way, the resolution and conversion gain is accurately known by the period of the clock which can be derived from a crystal oscillator [67].

2.4.5 Multi-Stage TDCs

In multi-stage TDCs, the time differences are converted in multiple stages. Architectures like coarse-fine and pipelined TDCs fit in this category. The main motivation for these architectures is to achieve a high dynamic range with good precision in the TDC.

2.4.5.1 Coarse-Fine TDCs

In coarse-fine TDCs, two TDCs are used. One which has a coarse delay line, followed by a second TDC which has a fine delay line [70]. The maximum delay which can be measured is set by the coarse delay line while the resolution is determined by the fine delay line. The coarse TDC makes the MSB decision. The residue is then processed by the fine TDC. Typically, a multiplexer is used to inject the residue from the coarse decision to the fine TDC. This multiplexer implies difficulties in terms of timing and matching [71]. Firstly, the delays require careful matching. Secondly, the coarse TDC has to make a decision fast to enable the multiplexer to the correct bin of the coarse line.

2.4.5.2 Pipelined TDCs

Pipelined TDCs rely on the amplification of the residue of each stage [72]. This is comparable to pipelined ADCs where each stage determines one or more bits and the residue is amplified by 2^n. Voltage amplification is implemented through amplifiers or switched capacitor circuits, but this is not trivial for time. To amplify the time difference between two intervals, the system has to rely on a circuit where the delay depends on the input time difference. Flip-flops or latches can implement this function [73]. When the time difference in a latch between Set and Reset is small, the circuit will approach its metastable state as is shown in Fig. 2.16. A linear

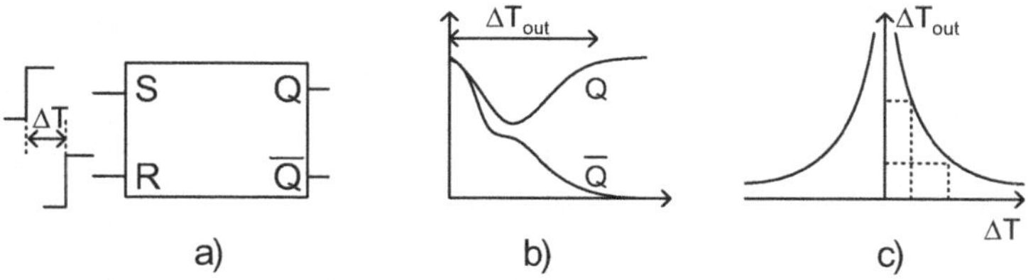

Fig. 2.16 (**a**) SR-latch used to amplify time differences. (**b**) Voltage waveforms in metastable region. (**c**) Output time-difference as function of the input time difference near the metastable region

Fig. 2.17 Pipelined TDC based on 1.5 bit MDAC TDCs which include a time-amplifier to multiply the time-difference by 2

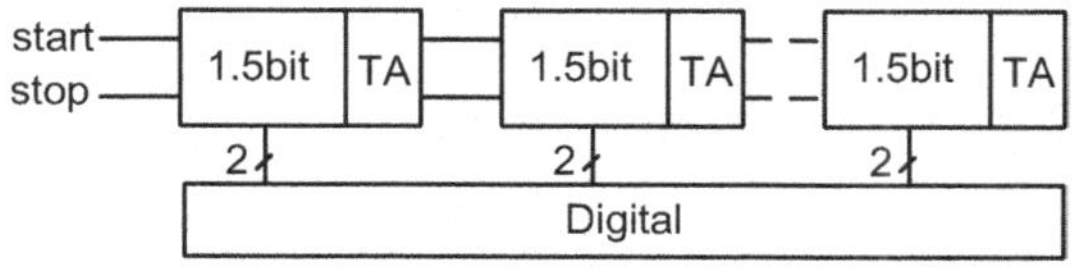

approximation can be done to model this circuit as a time difference amplifier. Real implementations will combine two time amplifiers, with certain time-offsets to linearize the region in between. The problem, however, with these time amplifiers is their robustness to process variation which is getting worse with ionizing radiation.

The time-amplifiers are typically used in a 1.5 MDAC (Multiplying Digital-to-Analog Converter) cell which has three output states $(-1, 0, +1)$. Compared to a 1-bit ADC, constraints on the offsets in each ADC are relaxed as there is an overlap between the codes which serves as an error correction. A typical pipelined architecture is shown in Fig. 2.17. Note that these time-amplifiers are also used in coarse-fine TDCs to improve the resolution of the TDC.

2.4.6 Looped TDCs

The dynamic range of the delay-line based architectures is limited by the length of the delay line. If the pulses arrive at the end, they cannot be observed anymore. A common practice is to implement the delay line as a ring, thus connecting the end back to the beginning [74]. When a hit arrives at the delay line, a pulse is injected in the line but this pulse is allowed to circle around the delay line. A counter will track the number of cycles the pulse has made in the delay loop. Due to imbalanced rise- and fall-times in the gates, pulses will start shrinking as they travel around. To ensure that the pulses do not vanish after some time, the pulses are regenerated each cycle with the pulse generator as is shown in Fig. 2.18.

This technique has been proven well to improve the dynamic range of delay-line based TDCs. The maximum dynamic range will be limited by pulse-shrinking effects, jitter, and non-linearity. Note that in this case, the mismatch between the cells are partially correlated since each cell is reused each cycle.

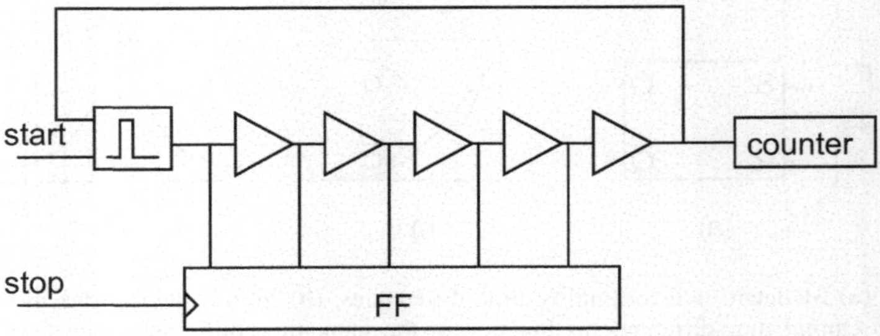

Fig. 2.18 Looped TDC architecture with pulse injection

2.4.7 Oversampling TDCs

All TDCs which have been discussed before are single-shot TDCs, this means they only require 1 time interval to make a measurement. Compared to ADCs, flash, SAR, and pipelined ADCs require only 1 voltage sample to do a measurement. Sigma-delta ADCs are oversampled converters and improve their resolution by oversampling and noise shaping. The same approach can be done in the time domain. This however implies that a repetitive time-interval is required at the input of the TDC. This is not common in high-energy physics applications where particles only occur once and single-shot TDCs are required. However, oversampling can be used in LIDAR applications since many identical measurements can be done.

When oversampling is used, the Signal-to-Noise ratio (SNR) due to quantization improves as quantization noise spreads equally over the frequency domain [75]:

$$SNR = 6.02N + 1.76 + 10\log(OSR) \tag{2.17}$$

where N is the amount of bits for each measurement and OSR is the oversampling ratio. Furthermore, the band above the Nyquist frequency is not used ($f > f_s/2OSR$) and can be used to push quantization noise with noise shaping.

2.4.7.1 Gated Ring-Oscillator

The architecture of a Gated Ring-Oscillator (GRO) is shown in Fig. 2.19 [76, 77]. The start and stop signals control a ring oscillator. Start enables the oscillator while stop disables the oscillator by switching the power and ground connections. The beauty of this circuit lies in the memory. When the GRO is disabled and restarted next time, ideally, the state (and phase) of the oscillator is remembered since all nodes remained floating in de oscillator.

If the quantization error of the start signal is ϵ_i and the quantization error for the stop signal is δ_i, the quantization error for a time interval measurement is $\delta_i - \epsilon_i$.

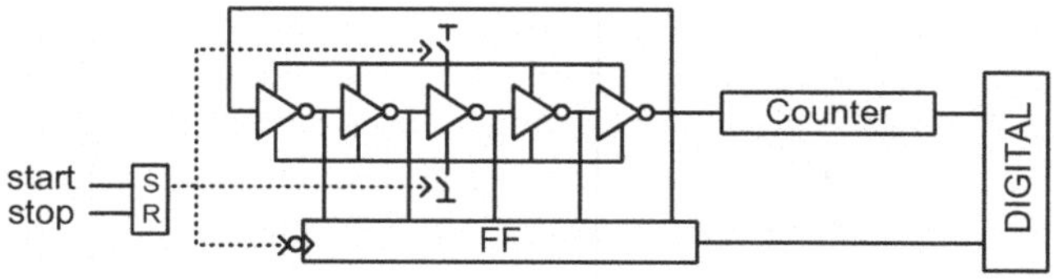

Fig. 2.19 Gated ring-oscillator

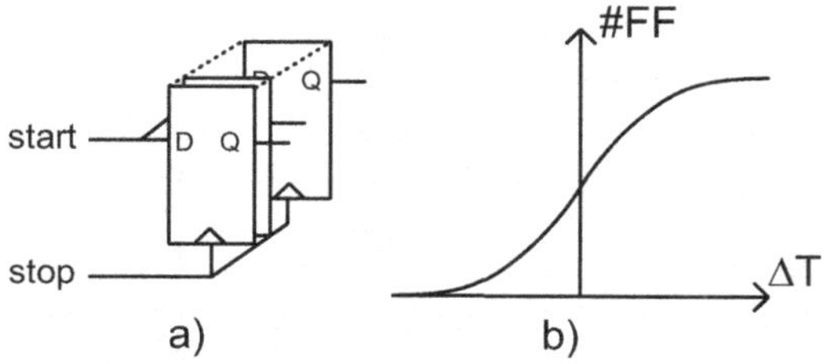

Fig. 2.20 (**a**) Stochastic TDC principle. (**b**) TDC transfer function based on the mismatch cumulative density function

The next cycle, the quantization error of the stop signal is still the same as where the oscillator had stopped on the stop signal. So, the quantization noise in a time-interval measurement is

$$\epsilon = \epsilon_{i+1} - \epsilon_i \tag{2.18}$$

This means that the quantization noise in the TDC is first order noise shaped, similar to a sigma-delta ADC.

Higher order noise shaping can be done with multi-stage noise shaping sigma-delta modulation (MASH) architectures in which the quantization noise is further noise shaped in multiple stages to higher order [78].

2.4.8 Other TDC Architectures

The architectures discussed above can be transformed towards other variants which improve the performance in a particular range. Some special architectures are particularly promising for nuclear applications due to their robustness.

2.4.8.1 Stochastic TDCs

All flip-flops in a circuit will intrinsically have local variations between other flip-flops on the same chip. These variations result in timing offsets on the decision point of the flip-flop. When a large array of flip-flops is placed with all data-pins connected to a start signal and all clocks connected to a stop signal, this timing mismatch will determine which flip-flops will show a 1 or 0 at the output as is shown in Fig. 2.20a. If the start signal arrives much earlier than the stop signal, all registers will be 1. Oppositely, if the start signal arrives very late, all registers will be 0. The transition between these regions is determined by the cumulative density function of the mismatch between the flip-flops, shown in Fig. 2.20b. As the flip-flop count increases, it is possible to very accurately read the time difference. Furthermore, this architecture is very error tolerant. If one flip-flop gets upset, for example due to radiation, it is only a small portion of the array and can be tolerated. This architecture can be used in combination with a delay line to improve its dynamic range [79].

2.4.8.2 Wave Union Launcher

Besides ASICs (Application Specific Integrated Circuits), FPGAs (Field Programmable Gate Arrays) are often preferred to implement digital functions due to their flexible hardware. FPGAs tend to be fabricated in advanced CMOS processes such as 16 and 12 nm CMOS technology. This implies that the logic in the FPGA becomes competitive to mid-range CMOS processes. If delay lines are carefully constrained in the FPGA, significant performance can be achieved, though some practical problems still occur.

An FPGA consists of an array of logic slices that can implement custom functions. Each slice contains a fast (mostly 4-bit) adder with carry-in and carry-out [80]. Dedicated inter-slice routing is used to interconnect different adders in the same row for high-end adders. These carry chains can be used to implement a fast delay line with bin sizes down to 20 ps.

The main problem in these architectures is linearity and extra-wide bins. The logic in the FPGA is not intended to be used as a matched delay line for TDCs. Unbalanced delays may occur but especially, extra-large delays may occur when crossing different clock domains or switching to a different column as the dedicated carry routes only travel from south to north on the FPGA.

The extra wide bins, however, can be digitally removed by means of a wave-union launcher as is shown in Fig. 2.21. A wave-union launcher injects a sequence into the delay line rather than a pulse. The idea behind this architecture is that, when the first edge of the sequence is within an extra wide bin, the other edges will not be in that region providing redundant measurements. Afterwards, this can be digitally compensated. Also, the circuit can be initially self-calibrated by histogram sampling to build a transfer function that can map DNL errors [81].

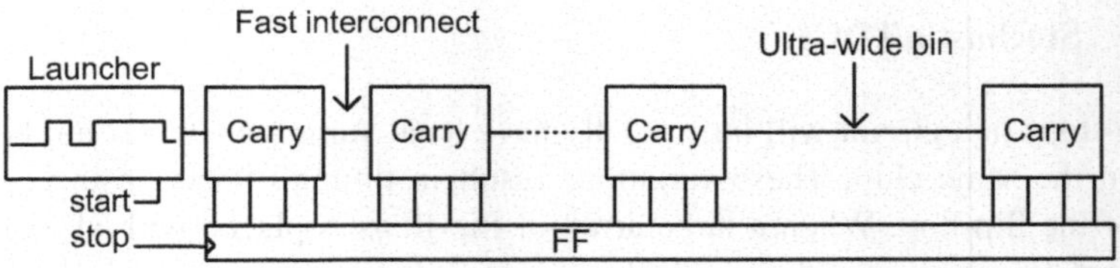

Fig. 2.21 Wave-union launcher

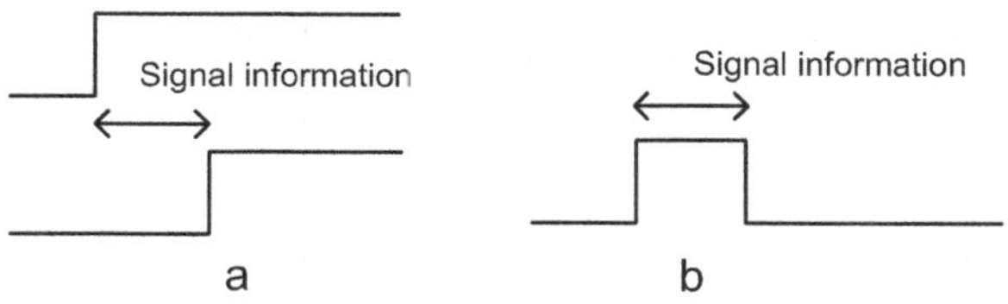

Fig. 2.22 (**a**) Time information as a delay between two signals. (**b**) Time information as pulse widths

The wave-union launcher has been widely used in FPGA based TDCs in high-energy physics experiments such as CERN and FermiLab where a resolution of several tens of picoseconds is sufficient.

2.4.9 Input Path

Some attention is required to the representation of the time interval. This can be done by two dependent signals in which the delay between these signals varies as shown in Fig. 2.22a. Another way is to only use one signal and modulate the pulse width (PWM = Pulse Width Modulation) of the signal. Both techniques have their advantages and disadvantages. Using two signals is the most straightforward way as a TDC typically accepts a start and stop signal. However, twice the power is also burned in the signal paths like input buffers and termination which may result in a significant increase of the power consumption of the system. In PWM signals, only one signal enters the system but care must be taken with the width of the pulse. First, a non-zero offset is needed since a minimal width is required for the input stages to process. Second, as the bandwidth of the input stages is limited, pulse width distortion occurs for narrow pulses leading to non-linearities.

Finally, the start/stop events can be synchronous or asynchronous. In the application scope of high-energy physics, the start signal is typically synchronous to a reference clock and can be seen as periodically. Therefore, in multichannel systems, only one reference is used for the multiple channels. In this way, the increased power consumption of the reference can be neglected.

2.5 Conclusion

An overview of different architectures for Time-to-Digital converts was given in this chapter. These circuits benefit from advances in digital speed for scaling CMOS processes. While basic TDCs are based on delay lines, advanced structures prove to work well for improving the resolution of the TDC beyond the minimum gate delay. Furthermore, oversampling can be used in the time-domain if periodic signals are available. Time-domain signal processing has become a promising way to implement systems though it requires an in-depth analysis of the timing.

Table 2.1 TDC architectures and their performance

	Resolution	Linearity	Power	Area	Radiation hardness (TID)
Flash	–	–	+	–	+
Vernier	+	–	–	–	–
Vernier looped	+	–	–	+	–
Parallel	+	–	–	–	–
Interpolation	+	+	+	+	+
Multi-stage	+	–	+	–	–
Oversampled	++	+	+	+	++
Stochastic	++	+	–	–	++

Nevertheless, it cannot be seen as a purely digital circuit. Table 2.1 shows a summary for the different TDC architectures which are discussed. Obviously, some architectures will prove their strengths depending on the application in which they are used.

Chapter 3
Clock Synthesizers

3.1 Introduction

Frequency synthesizers comprise an important range of circuits in many SoCs (Systems-on-Chips). A synthesizer will create a frequency based on a different reference clock. These clock synthesizers are used in digital systems to generate high speed clocks, based on a low frequency crystal oscillator. Synthesizers are further used to generate the local oscillator (LO) signal in RF communication systems to upconvert or downconvert the base-band or RF signal. Third, for a special range of circuits, called Clock-Data recovery (CDR) circuits, the frequency reference is a data-stream rather than a periodic clock. The circuit has to recover the clock signals from the serial communication data input.

Clock synthesizers are in general Phase-Locked Loops (PLLs). PLLs exist in many types, frequency ranges and targeted applications. Some circuits target a diverse frequency-band while others operate only at a single frequency. The trade-offs, complexity, and noise levels depend on the PLL architecture and its implementation which will be discussed in this chapter.

3.2 Phase Locked Loops

A PLL consists in its basic form of four components: a phase detector (PD), loop filter (LF), voltage controlled oscillator (VCO), and divider. The main philosophy in PLLs is that the VCO generates an output clock which is after division compared to a reference clock by the phase detector. The VCO signal phase is adjusted by theloop until the phase difference between the reference and the feedback is zero.

J. Prinzie et al., *Radiation Hardened CMOS Integrated Circuits for Time-Based Signal Processing*, Analog Circuits and Signal Processing,
https://doi.org/10.1007/978-3-319-78616-2_3

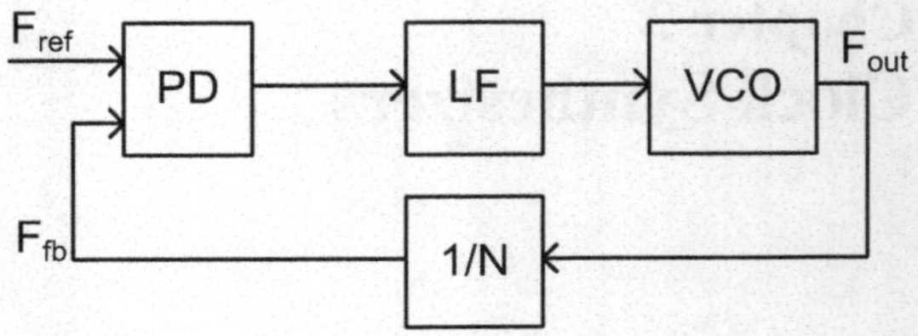

Fig. 3.1 PLL block diagram with a phase detector (PD), loop filter (LF), voltage controlled oscillator (VCO), and divider (1/N)

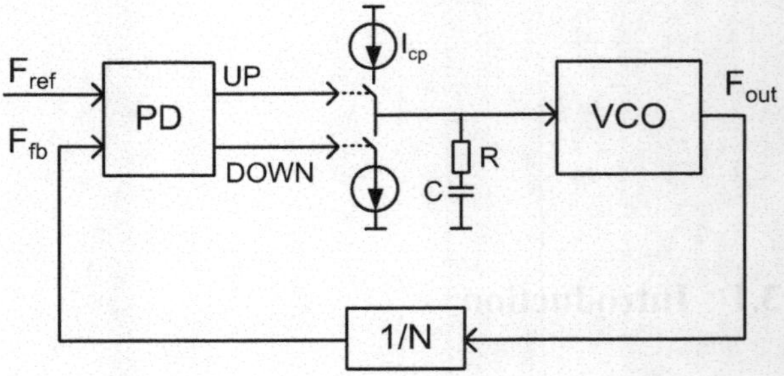

Fig. 3.2 PLL block diagram with charge-pump and passive loop filter

The PLL is then said to be locked. A block diagram of a PLL is shown in Fig. 3.1. The loop will ensure that the phase of the reference clock (f_{ref}) is aligned to the phase of the feedback clock (f_{fb}).

If the phase between the two clocks is locked, then

$$\phi_{ref} = \phi_{fb} \rightarrow \frac{d\phi_{ref}}{dt} = \frac{d\phi_{fb}}{dt} \rightarrow \omega_{ref} = \omega_{fb} \tag{3.1}$$

Phase lock also implies frequency lock. This ensures that the output frequency of this structure is $f_{out} = N \cdot f_{ref}$. Circuits implemented by this scheme are called integer-N frequency synthesizers as they can upscale the frequency only by an integer number. In contrast, fractional-N synthesizers can synthesize intermediate frequencies between N and $N + 1$ by toggling between two division rates [82].

In charge-pump based PLLs, the loop filter is implemented as a charge-pump which converts the phase detector signals into a current. The phase detector will generate up- and down-signals with a width proportional to the phase difference as shown in Fig. 3.2.

A passive or an active filter [83] converts the charge-pump current into a control voltage. In a passive filter, the current which is injected in the capacitor acts as an integrating path, while the voltage drop, generated by the resistor acts as a proportional path. The combination of both is an integrator with a zero at $\omega = 1/RC$.

3.2.1 Phase-Domain Model

The dynamics of a PLL are analyzed in the phase domain as is shown in Fig. 3.3. The phase domain model of a phase detector is simply a subtracter in which K_{pd} represents the gain of the phase detector which is typically $1/(2\pi)$. The charge-pump current is represented as I_{cp}. A divider can be represented in the phase domain by a gain block of $1/N$ since the frequency is divided by N and thus any time difference in the clock is referenced to a smaller frequency. The loop filter can be represented by a frequency domain transfer function ($Z(s)$). Finally, the VCO can be represented as an integrator since

$$\phi = \int \omega dt = \int K_{vco} V_{in} dt \tag{3.2}$$

in which K_{vco} is the voltage-to-frequency gain of the oscillator ((rad/s)/ V) and V_{in} its control voltage.

This feedback loop has an open loop transfer function of

$$H_{ol} = \frac{1}{2\pi} Z(s) \frac{K_{vco}}{s} \frac{1}{N} \tag{3.3}$$

The filter $Z(s)$, according to Fig. 3.2, is

$$Z(s) = \frac{1 + s \cdot RC}{sC} \tag{3.4}$$

The closed loop transfer function of the PLL is

$$H_{cl}(s) = \frac{\phi_{out}}{\phi_{ref}} = \frac{\frac{1}{2\pi}\frac{I_{cp}}{C}\frac{1+sRC}{s}\frac{K_{vco}}{s}}{1 + \frac{1}{2\pi}\frac{I_{cp}}{C}\frac{1+sRC}{s}\frac{K_{vco}}{s}\frac{1}{N}}$$

$$= \frac{\frac{I_{cp}K_{vco}}{2\pi C}(1+sRC)}{s^2 + s\frac{1}{N}\frac{I_{cp}K_{vco}}{2\pi}R + \frac{1}{N}\frac{I_{cp}K_{vco}}{2\pi C}}$$

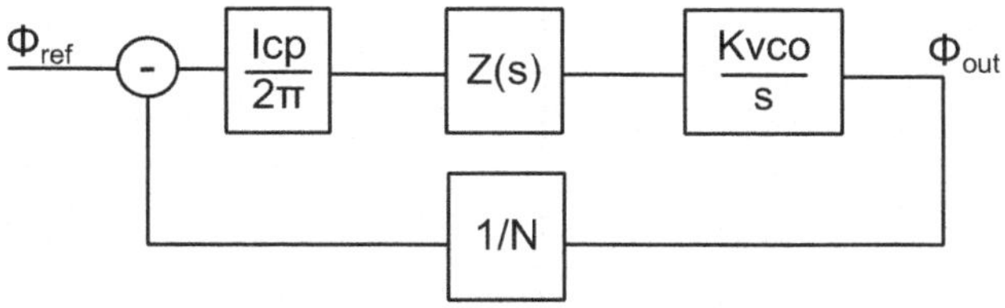

Fig. 3.3 Phase domain model of a PLL

This can be rearranged to a classical second order transfer function [84]:

$$N\frac{\omega_n^2(1+s/\omega_z)}{s^2+2\zeta\omega_n+\omega_n^2}=N\frac{\omega_n^2(1+\frac{2\zeta}{\omega_n}s)}{s^2+2\zeta\omega_n+\omega_n^2} \tag{3.5}$$

in which

$$\omega_z=\frac{\omega_n}{2\zeta}=\frac{1}{RC}$$

$$\omega_n=\sqrt{\frac{1}{N}\frac{I_{cp}K_{vco}}{2\pi C}}$$

$$\zeta=\frac{R}{2}\sqrt{\frac{1}{N}\frac{I_{cp}K_{vco}C}{2\pi}}$$

First of all, the closed loop transfer function is a low pass filter which implies that fast phase variations (or noise) are filtered by the loop and slow variations are tracked. The loop stability is important since an under-damped system will increase the noise and acquisition time of the PLL. The stability can be easily analyzed by the open loop transfer function or the parameters from (3.5). Figure 3.4 shows the open loop transfer function in magnitude and phase. The loop is stable if the phase margin at the crossover frequency is sufficiently large, typically more than 60°. The plot shows that, if the open loop gain is increased, the phase margin increases as well. Also, if the frequency of the zero is lowered, the stability also increases.

The magnitude from (3.5) can be used to calculate the bandwidth of the loop. This reduces to

$$\omega_{-3\,\text{dB}}=\sqrt{1+2\zeta^2+\sqrt{(1+2\zeta^2)^2+1}}\omega_n \tag{3.6}$$

If ζ is between $\sqrt{2}/2$ and 1, $\omega_{-3\,\text{dB}}$ varies from $2.1\omega_n$ to $2.5\omega_n$. If ζ is large, $\omega_{-3\,\text{dB}}$ approaches $2\zeta\omega_n$ which can be explained from a single pole estimation.

The typical design procedure for a PLL loop is as follows. Given a certain bandwidth, the damping factor is chosen such that the loop is sufficiently damped.

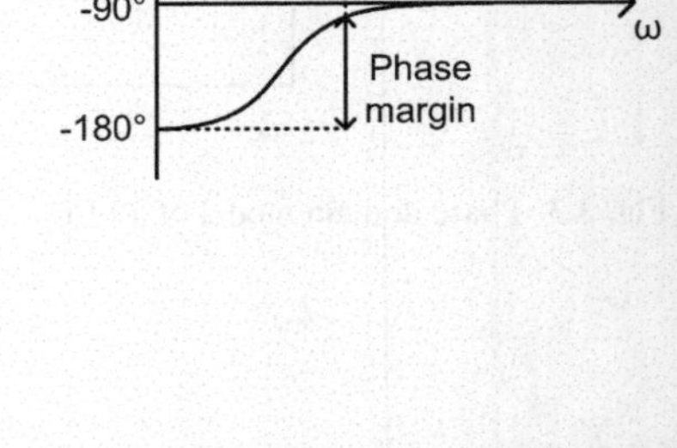

Fig. 3.4 Open loop transfer function of a second order PLL

As an example, $\zeta = 1$. From (3.6), ω_n can be calculated. For $\zeta = 1$, $\omega_n = \omega_{-3\,\mathrm{dB}}/2.5$. From ω_n, ω_z can be calculated being $\omega_z = \omega_n/(2\zeta)$. In most designs, K_{vco} and N are fixed. The loop parameters are then set with I_{cp}, R and C. The equation has 2 degrees of freedom, set by 3 parameters. Given ω_n, the ratio I_{cp}/C is known. The value of I_{cp} or C can be chosen, usually depending on noise criteria. This will be discussed further. The zero is then set by 1/RC.

None of the parameters R, C, or I_{cp} can independently tune the bandwidth without altering the damping of the loop. In many designs, both I_{cp} and R are made configurable to adjust the bandwidth with a constant damping in the loop.

3.2.2 Components for Charge-Pump PLLs

3.2.2.1 Phase Detector

Phase detectors calculate the phase difference between the reference clock and feedback clock. They are simply modeled as a subtracting element since the signal, produced by these circuits is proportional to the phase difference. Phase detectors exist in many forms. In the past, analog PLLs were constructed based on mixers that serve as a phase detector. These implementations generate an error signal of

$$V_{pd} = A_{ref}A_{fb}\sin(\phi_{ref} - \phi_{fb}) \approx A_{ref}A_{fb}(\phi_{ref} - \phi_{fb}) \tag{3.7}$$

in which $A_{ref}/\phi ref$ and $A_{fb}/\phi fb$ are the amplitude/phase of the reference and feedback signal, respectively. The problem with this approach is the dependency on the signal amplitude which determines the gain of the phase detector.

The majority of the PLLs today use digital logic to implement phase detectors: Phase-frequency detectors (PFDs) [85] and Bang-Bang phase detectors [86, 87].

3.2.2.2 Phase-Frequency Detector

A Phase-Frequency Detector is shown in Fig. 3.5 which acts as a sequential system, implemented by asynchronous reset flip-flops. For example, when the reference clock rises, the up signal rises and remains high until the feedback clock rises. This invokes a reset on both flip-flops. Figure 3.6a shows an example of the PFD signals when the reference clock is leading the feedback clock. The time when both up and down signals are high is T_{reset}.

In PLLs, the up and down signals control the currents of the charge-pump in the loop. This means that the average output value of (up-down) determines the error signal of the phase detector. In this example, the up signal is high for a time

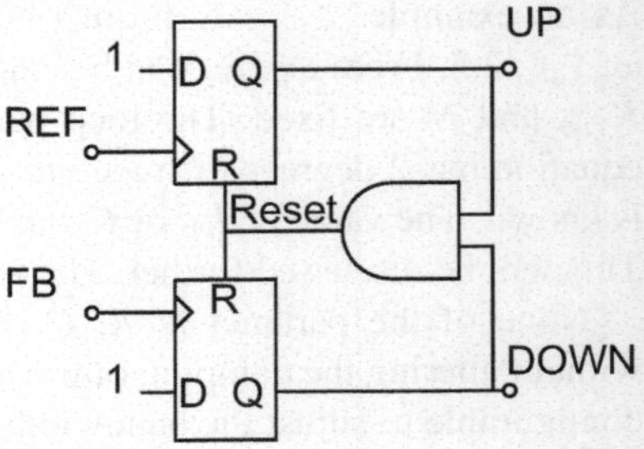

Fig. 3.5 Phase-frequency detector logic

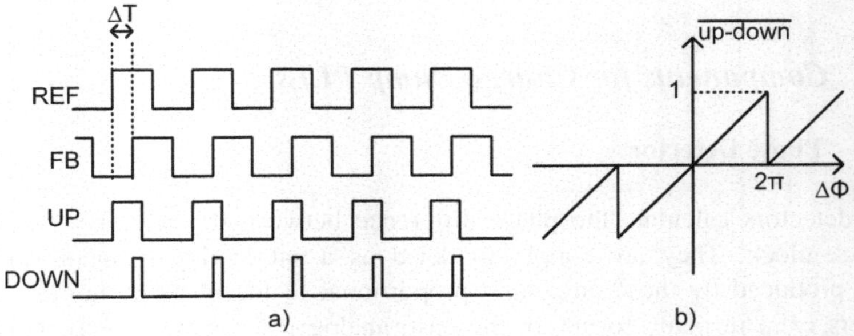

Fig. 3.6 (**a**) Signals at the PFD. (**b**) PFD transfer function

$$T_{up} = (\phi_{ref} - \phi_{fb})\frac{T_{ref}}{2\pi} + T_{reset} \tag{3.8}$$

while the down signal is high for only

$$T_{down} = T_{reset} \tag{3.9}$$

The average output signal of the difference between up and down is then

$$\langle up - down \rangle = \frac{T_{up} - T_{down}}{T_{ref}} = \frac{1}{2\pi} \cdot \Delta\phi \tag{3.10}$$

The transfer function is highly linear as is shown in Fig. 3.6b.

The time for which both outputs are on depends on the delay in the reset path. This delay is essential and should be sufficient to avoid dead-zone. Dead-zone occurs when small phase differences do not generate corrections to the loop. If, for example, the flip-flops are reset too early such that the up and down pulses become too narrow, signal buffering to the charge pump may fail or the charge pump may not have sufficient time to push the required charges to the filter. This effect, if present, will contribute to significant amounts of jitter since small phase noise deviations will not be compensated by the loop [88].

Besides a phase comparison, PFDs also provide frequency detection. If the reference frequency is higher than the feedback frequency, there are on average more clock edges on the reference clock resulting in an average higher up value. Though the frequency detection is non-linear, it works for infinite frequency difference. This makes PFDs an attractive implementation since the pull-in range (the maximum initial frequency deviation for which the PLL is still able to lock) is infinite.

3.2.2.3 Bang-Bang Phase Detector

Bang-Bang phase detectors only involve a single flip-flop which acts as an early or late detection being a highly non-linear detector. In contrast to PFDs, Bang-Bang phase detectors do not have infinite pull-in range [89] but can easily operate at higher frequencies and are more feasible for CDR applications and All-Digital PLLs.

3.2.2.4 Charge-Pump: Loop Filter

Charge-pump circuits are able to source or sink a current, depending on its input. The concept of a charge-pump is shown in Fig. 3.7a. Figure 3.7b shows an example of the integration of a charge-pump current on a capacitor. The combination PFD and charge-pump can thus be seen as a gain of $I_{cp}/2\pi[A/\text{rad}]$ which converts a phase difference to a current which flows in the loop filter impedance. The output voltage at the charge-pump can be generalized as

$$V_{cp(s)} = \frac{I_{cp}}{2\pi} Z(s)(\phi_{ref}(s) - \phi_{fb}(s)) \tag{3.11}$$

in which $Z(s)$ is the loop filter impedance. In case of an RC series network, the transfer function becomes

$$\frac{V_{cp}(s)}{\phi_{ref}(s) - \phi_{fb}(s)} = \frac{I_{cp}}{2\pi}\frac{1+sRC}{sC} \tag{3.12}$$

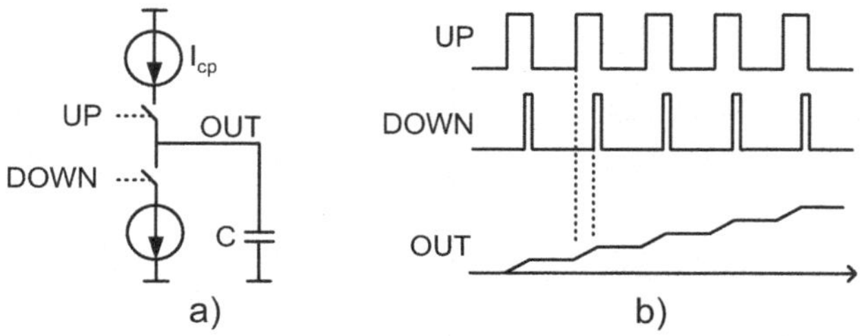

Fig. 3.7 (**a**) Conceptual charge pump. (**b**) Example of the output charge integration on a capacitor

Ideally, when both up and down would be high, the charge pump would not source or sink any current. Each cycle, the voltage is increased by

$$\Delta V = \frac{\Delta\phi}{2\pi}\frac{I_{cp}}{C}T_{ref} \tag{3.13}$$

with an average slope of

$$\frac{\Delta V}{T_{ref}} = \frac{\Delta\phi}{2\pi}\frac{I_{cp}}{C} \tag{3.14}$$

The circuit in Fig. 3.7 and its analysis assumes that the sinking and sourcing current is the same. However, in practice, there will be an imbalance between the up and down currents. This leads to two effects in the PLL: static phase offset (SPO) and spurs [90].

If the PLL is in lock, the net current to the loop capacitor is zero so the average source and sink current must be equal. If, for example, the up is smaller than the down current, the PLL will introduce a phase offset between the reference and feedback clock to ensure that

$$(T_{spo} + T_{reset})I_{up} = (T_{reset})I_{down}$$
$$\rightarrow T_{spo} = \frac{\Delta I}{I_{up}} \cdot T_{reset}$$

Typical values for T_{reset} are 100 ps up to 400 ps. With a 5% current mismatch, this leads to 5 ps up to 20 ps offset.

The output current waveform will be a pulsed waveform I_{up} for $[0 : T_{spo}]$ and ΔI for $[T_{spo} : T_{spo} + T_{reset}]$. Repeated at f_{ref}.

This current spike will create voltage spikes across the resistor in the loop filter.

The voltage at the output of the loop filter is a function

$$V_{spur}(t) = \sum_{n=1}^{\infty} 2|c_n|\cos(2\pi n \cdot \omega_{ref} + \arg(c_n)) \tag{3.15}$$

in which c_n are the Fourier series components and ω_{ref} is the reference clock's pulsation.

This signal consists of spurious tones at frequencies nf_{ref}. The output spectrum of the phase is further shaped by the integration of the VCO :

$$\phi_{spur}(f) = \sum_{n=-\infty}^{\infty} c_n \cdot \delta(f - nf_{ref}) \cdot \frac{K_{vco}}{n\omega_{ref}} \tag{3.16}$$

These spurs can be reduced in a few ways. The first way is to optimize the charge pump and to reduce the current imbalance [91, 92]. The second approach is to reduce the value of K_{vco}. A third solution is to implement a higher order loop as is shown in Fig. 3.8a. The capacitor $C2$ filters the pulses generated over the resistor which reduces the reference spurs. Note that this capacitor will create an open-loop pole at

$$\omega_p = \frac{1}{R\frac{C_1+C_2}{C_1C_2}} \approx \frac{1}{RC_2} \tag{3.17}$$

This pole may reduce the phase margin and the stability of the loop. Ideally, the PLL loop is designed such that the crossover frequency is the logarithmic center of ω_z and ω_p to maximize the phase margin. The larger ω_p/ω_z, the higher the maximum phase margin can be. The open loop transfer function is shown in Fig. 3.8b.

The current sources can be implemented as shown in Fig. 3.9a with their biasing circuit. A current DAC is used to set the current of the charge-pump which is mirrored to the output devices. This practical implementation however has some limitations. Firstly, the output devices do not have the same drain voltages as the

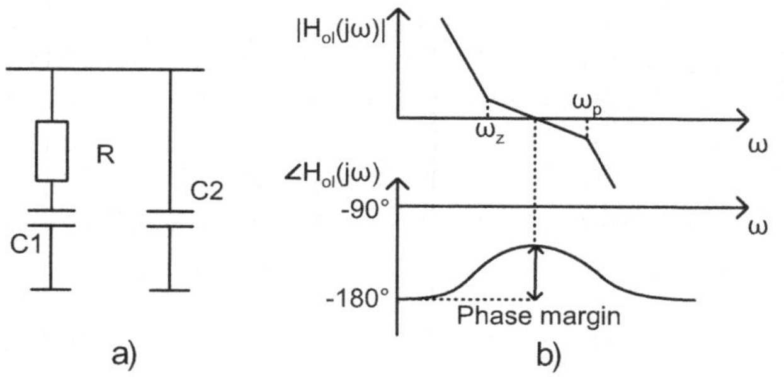

Fig. 3.8 (**a**) Second order loop filter to reduce reference spurs. (**b**) Open loop PLL bandwidth with second order loop filter

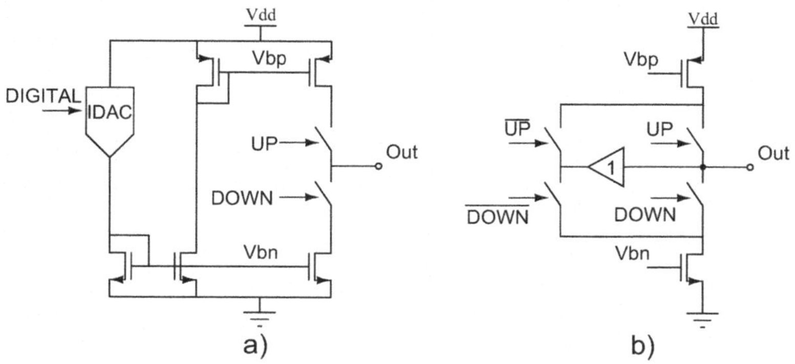

Fig. 3.9 (**a**) CMOS charge pump implementation and biasing circuit. (**b**) Dual path that provides a constant flow of current

mirror devices since $V_{out} \neq V_{bp}$ and $V_{out} \neq V_{bn}$. This leads to a systematic DC imbalance between up and down currents. Secondly, the output impedance at the output of the nmos and pmos devices results in an imbalance between both currents. When the charge-pump is designed, the currents are matched when the output voltage is ≈Vdd/2. However, when the PLL is in lock, the voltage may not be this predetermined value. The up and down currents diverge when the output voltage is further from its midpoint due to the output characteristics of the devices. When the output approaches $V_{gs} - V_T$ (for the nmos device), the nmos transistor enters the triode region while the pmos is in deep saturation leading to a severe current imbalance.

To overcome this, the output voltage swing can be limited which limits the current imbalance. An active loop filter with a virtual ground at the input can ensure that the charge-pump's output node remains constant. However, additional noise is introduced by the active elements of the filter. Secondly, the tuning limits of the VCO can be limited by adding additional bands to the oscillator such that wide frequency corrections are done initially by a digital state-machine [93] and small (noise) corrections are done by the PLL loop [94].

The output impedance of the charge pump can be improved with cascode devices [92] or source degeneration [95, 96]. Cascode devices reduce the voltage headroom with V_{gst}. This has proven to work well but the main drawback is additional circuitry to generate the bias voltages which are prone to radiation effects as the headroom reduces with dose. A simple and effective approach to boost the impedance is source degeneration as is shown in Fig. 3.10. In this case the output resistance is

$$R_{out} = R_{degen} + r_{ds}(1 + g_m R_{degen}) \tag{3.18}$$

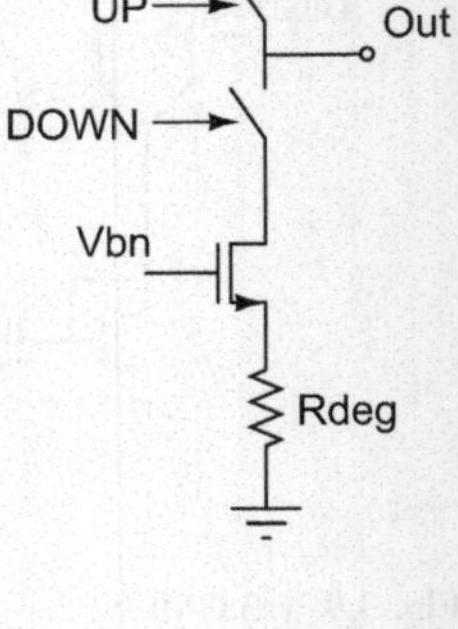

Fig. 3.10 Source degenerated current sources with improved output resistance

The output impedance will still be determined by the headroom which is needed for the design. In this case

$$1 + g_m R_{degen} = 1 + \frac{2 I_d R_{degen}}{V_{gst}} = 1 + \frac{2 V_{drop}}{V_{gst}} \tag{3.19}$$

in which V_{drop} is the additional voltage drop on the degeneration resistance. Generally, V_{drop} is not exceeding V_{gst} to save headroom. When $V_{gst} = V_{drop}$, the output resistance is boosted by a factor of 3 which may be sufficient.

Another problem arises from its transient behavior. When the charge-pump is off, no current is flowing and the drain voltages of the nmos and pmos current sources will end at the ground and supply, respectively. At the moment when the up or down switch is activated, charge sharing occurs between the output capacitance and the total capacitance seen at the drain of the current sources. Charge sharing can be reduced with a stand-by branch as is shown in Fig. 3.9b which ensures that the current can continue to flow when the output switches are disabled. A unity gain amplifier tracks the output voltage to the stand-by branch. Different architectures were proposed to solve the issues of current matching and charge sharing in [97] and [98].

3.2.2.5 Divider

Dividers may come in different forms, depending on the operation frequency. Typically, frequencies up to 3–4 GHz can be processed with standard cell CMOS logic. At higher frequencies, a prescaler implemented with Current Mode Logic (CML) can reduce the frequency until the logic is fast enough to divide further in CMOS logic. CML logic consumes more power but operates at higher speeds. Divide-by-2 CML gates are usually implemented as master–slave latches [99].

The majority of CMOS dividers are implemented with digital standard cells and can be fully synthesized. Either the design is done with a ripple counter strategy (Fig. 3.11a) or fully synchronously (Fig. 3.11b). The advantage of the first approach is that only the first gate operates at the input rate which can save power. However, jitter which originates in the logic is cascaded and accumulates through the divider. This is not the case with synchronous logic where the jitter is only determined by one flip-flop. This topology fully runs at the input rate. Synchronous logic can also be better hardened against single-event upsets using TMR which is less straightforward with ripple counters.

3.3 Oscillators

Oscillators are circuits which produce a periodic signal without any input. For PLLs, voltage controlled oscillators are required since the frequency of the oscillator needs

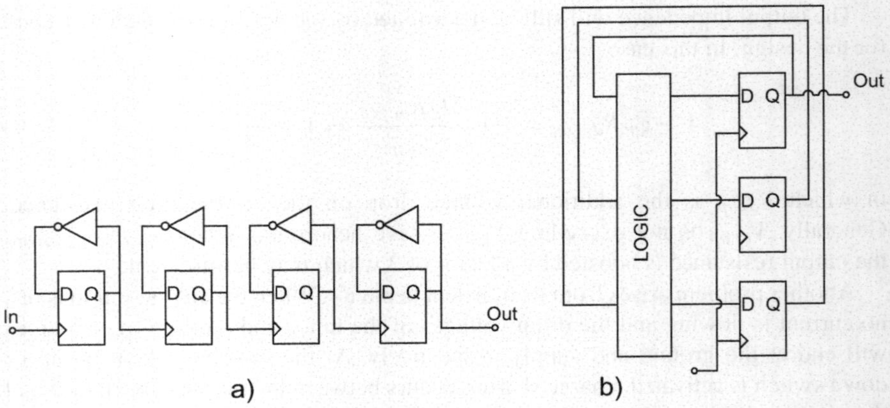

Fig. 3.11 (**a**) Ripple counter based divider. (**b**) Synchronous logic divider implementation

to be controlled by the loop of the PLL. The output voltage signal of the oscillator can be described as

$$V(t) = A \cdot f\left(\int \omega(t)dt + \phi_n(t)\right) = A \cdot f\left(\int (\omega_0 + K_{vco}V_{tune}(t))dt + \phi_n(t)\right) \quad (3.20)$$

in which A is the oscillation amplitude, $f(\cdot)$ is a periodic function with a period of 2π, ω_0 is the oscillation center pulsation, $V_{tune}(t)$ its control voltage, and $\phi_n(t)$ the phase noise expressed in the time domain.

3.3.1 Oscillation Criteria

An oscillator relies on positive feedback for a signal to grow in the circuit. The criteria for an oscillation to start is that the circuit has complex poles in the right-half plane, such that the time-domain impulse response is

$$h(t) = e^{\alpha t}(A \cdot e^{\omega_0 t} + B \cdot e^{-\omega_0 t}) \quad (3.21)$$

with $\alpha > 0$ rad/s. Any signal which exists in the circuit (e.g., noise) triggers the growing oscillation. When the oscillation is stable, the amplitude is constant which means that $\alpha = 0$ and the poles lie on the imaginary axis.

This leads to the Barkhausen criterion for an oscillator [100]:

$$|H_{ol}(\omega_0)| = 1$$
$$\arg(H_{ol}(\omega_0)) = 2\pi n, n = [0, 1, 2, \ldots]$$

which means that the total loop gain in the circuit is 1 and the phase shift is 360°. Practical oscillators are however designed such that $|H_{ol}| > 1$ which results in an increasing amplitude over time. However, due to non-linearities of the large signals, the gain of the oscillator will drop at certain amplitudes until it reaches one.

3.3.2 LC-Oscillators

LC-oscillators depend on the resonance of an LC-tank. The general form of the oscillator can be seen in Fig. 3.12 which includes a parallel (non-ideal) LC tank and a negative resistance $-R$ which compensates for the losses in the tank [101].

Each passive element, like an inductor or capacitor, introduces losses. The inductor, for example, has an equivalent series resistance due to the substrate losses, eddy currents, and resistive losses. While this equivalent model is simplified, it proves good insight in the trade-offs of the components [102].

At a given frequency, the series resistance can be converted to a parallel resistance and a parallel reactive component:

$$L_p = L_s \frac{Q^2+1}{Q^2} \approx L_s$$

$$C_p = C_s \frac{Q^2}{Q^2+1} \approx C_s$$

$$R_p = R_s(Q^2+1) \approx Q^2 R_s$$

$$Q_L = \frac{\omega_0 L_s}{R_s} = \frac{R_p}{\omega_0 L_p}$$

$$Q_C = \frac{1}{\omega_0 C_s R_s} = \omega_0 C_p R_p$$

If the quality factors (Q_L and Q_c) are sufficiently large, the equivalent series and parallel reactive components are approximately equal. The quality factor is defined as the ratio between the energy stored and the energy lost in each cycle. If multiple elements are combined, the total quality factor is

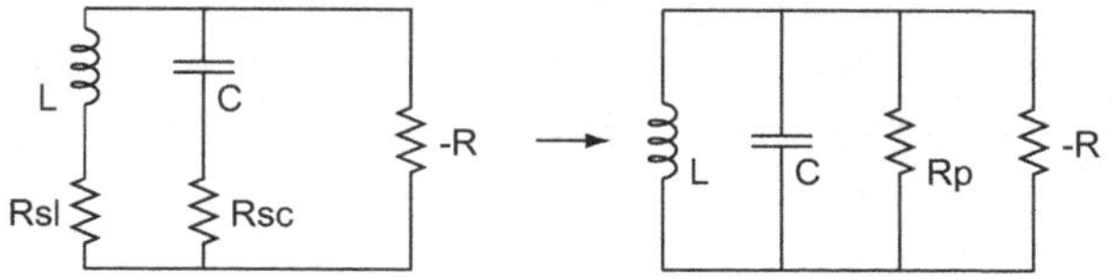

Fig. 3.12 Parallel resonance LC tank with series and equivalent parallel resistance

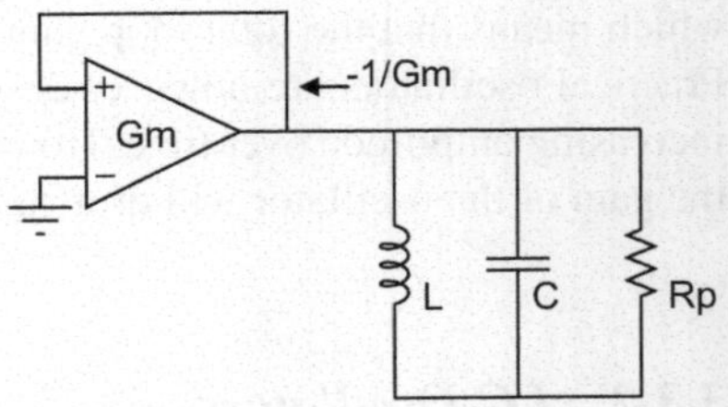

Fig. 3.13 Parallel resonance LC tank with $-1/G_m$ parallel negative resistance implemented with positive feedback

$$\frac{1}{Q_{tot}} = \sum \frac{1}{Q_n} \tag{3.22}$$

The tank will resonate at a frequency

$$\omega_0 = \frac{1}{\sqrt{L_p C_p}} \tag{3.23}$$

when the parallel reactive impedance is infinite. When the tank resonates, the total parallel resistance seen in the tank is the total R_p:

$$R_p = Q_{tot}\omega_0 L_p = \frac{Q_{tot}}{\omega_0 C_p} \tag{3.24}$$

The negative resistance in Fig. 3.12 can be implemented as shown in Fig. 3.13. An amplifier with transconductance of G_m is applied with positive feedback. This leads to a negative resistance of $-1/G_m$. The oscillation condition for this circuit is satisfied if

$$G_m R_{p-tot} > 1 \tag{3.25}$$

This can also be explained by the total loop gain which is $G_m R_p$. This implementation however does not have any amplitude control.

A real circuit implementation is shown in Fig. 3.14a. The negative resistances are implemented with cross coupled differential pairs which provide a negative resistance of $-2/g_m$, both for the pmos and nmos stages (Fig. 3.14b). When the loop gain is sufficient in this circuit, such that the cross coupled pairs provide full switching, the amplitude of the oscillation can be determined by the biasing current. In this case, a square wave current flows through the tank with an amplitude of I_{bias}. This square wave, flowing through the resonant tank will be filtered such that mainly the first harmonic is converted to a voltage. The (differential) voltage amplitude across the tank is then

$$V_d = \frac{4}{\pi} R_p I_{bias} \tag{3.26}$$

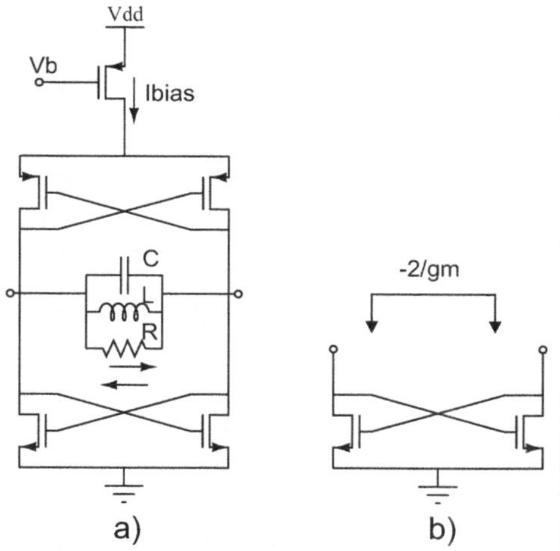

Fig. 3.14 (**a**) CMOS integrated LC oscillator. (**b**) Equivalent negative resistance of a cross coupled pair

or a single-ended voltage amplitude at both nodes of $\frac{2}{\pi} R_p I_{bias}$.

The oscillation frequency of this circuit can be adjusted by tuning the value of the capacitor using varactors. A detailed discussion on this tuning topology is found in Chap. 5.

3.3.3 Ring-Oscillators

Ring oscillators are digital-like oscillators which are implemented as a chain of delay cells, similar to the delay lines in TDCs. If the number of delay cells is odd, a ring oscillator with N inverters oscillates at a frequency

$$f_0 = \frac{1}{2NT_{cell}} \tag{3.27}$$

If an even number of cells is used, the oscillator has a stable state for f=0, being DC. Compared to integrated LC oscillators, ring oscillators occupy much smaller silicon area since the inductor and capacitors are omitted. To ensure the ring oscillators start, sufficient gain is required in each cell. The delay blocks can be modeled as a single pole amplifier:

$$H(j\omega) = A\frac{1}{1 + j\omega/\omega_{-3\,\mathrm{dB}}} \tag{3.28}$$

for which the total phase shift in each cell is π/N such that the circuit starts its oscillation at $\omega = \tan(\pi/N)\omega_{-3\,\mathrm{dB}}$ [103]. At this frequency, the total gain

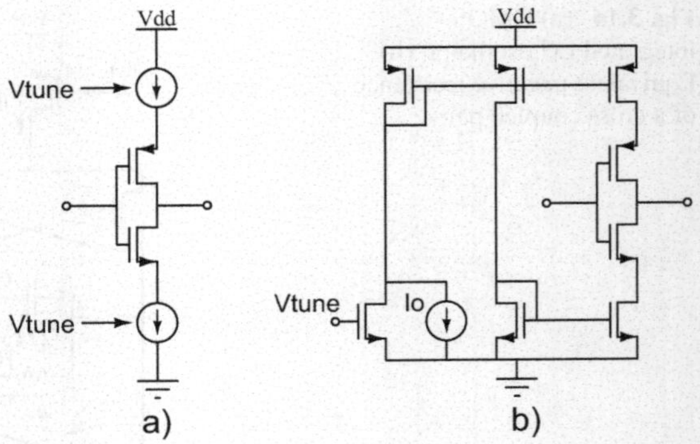

Fig. 3.15 (**a**) Current starved delay cell. (**b**) Circuit implementation of a current starved delay cell

$$A\frac{1}{1 + j \cdot \tan(\pi/N)} \tag{3.29}$$

should be larger than 1. For 3 cells, the gain reduction is 2 while for 9 cells, this is only 1.06.

The frequency of the oscillator is tuned by altering the delay of the cells. Single ended circuits can be current limited and are so called "current starved" delay cells to tune the frequency. A circuit is shown in Fig. 3.15a. Both bias currents are derived from a current mirror from which the current is adjustable through a voltage source (Fig. 3.15b). The delay of this cell can be approximated as

$$T_{cell} = \frac{V_{dd}C}{I(V_{tune})} \tag{3.30}$$

A zero-voltage current I_0 is required to ensure a minimum frequency when the input voltage is below the threshold voltage of the input transistor.

The current sources can also be shared by all delay cells. In this way, a current controlled oscillator (CCO) is formed [104].

Many circuit implementations involve differential delay cells. Firstly, these cells can be implemented with an even or odd number of delay cells and secondly, they are less prone to supply noise compared to single-ended cells. In terms of jitter, as discussed on delay lines in Chap. 2, single-ended circuits require less average power for the same jitter performance compared to differential circuits [63]. An example of a CMOS delay cell is shown in Fig. 3.16a. The delay of these differential circuits is proportional to $\tau = R_{out}C_L$. The delay is adjusted with a triode load of which the resistance is adjustable. While tuning, the current source is adjusted proportionally to keep the common mode output voltage and differential amplitude constant. The biasing circuit in Fig. 3.16b is an example of this implementation where the swing is set by V_{set}.

A ring oscillator was designed as a comparison to an integrated LC-oscillator which is discussed in Chap. 5.

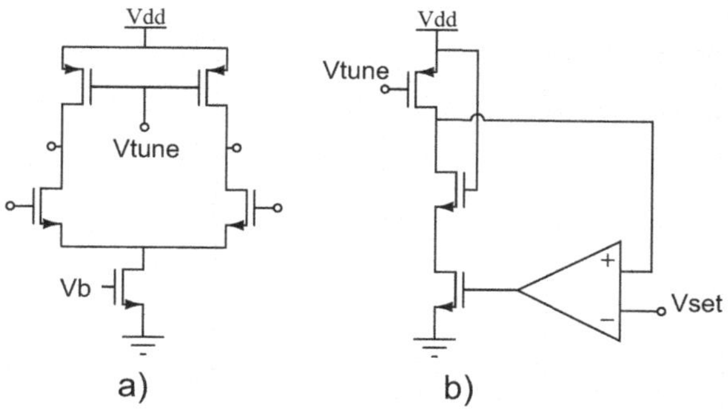

Fig. 3.16 (**a**) Differential delay cell. (**b**) Biasing circuit for constant amplitudes

3.4 Jitter and Phase-Noise

3.4.1 Definitions

Phase noise refers the frequency power spectral density of the phase fluctuations $\phi_n(t)$ in

$$V(t) = A \cdot \sin(\omega_0 t + \phi_n(t)) \tag{3.31}$$

This power spectral density is expressed as $S_\phi(f)$ and is defined as

$$S_\phi(f) = \mathcal{F}\{R_{\phi\phi}(t)\} \tag{3.32}$$

in which $R_{\phi\phi}(t)$ is the autocorrelation of the phase domain noise.

Single-SideBand (SSB) voltage noise is closely related to this phase noise. When the voltage spectrum of a noisy oscillator is observed, it is not an impulse at f_0 but it contains noisy skirts along its center frequency as is shown in Fig. 3.17 [105]. The SSB noise is defined as

$$\mathcal{L}(\Delta f) = \frac{P_{sideband}(f_0 + \Delta f, 1\,\text{Hz})}{P_{sig}} \tag{3.33}$$

in which P_{sig} is the power of the carrier and $P_{sideband}(f_0 + \Delta f, 1\,\text{Hz})$ is the noise power in a 1 Hz band. This SSB is expressed in dBc/Hz since it is the noise power related to the carrier power. Note that SSB noise includes all types of noise: AM and PM, though, phase noise only expresses the noise from phase variations. Since most oscillators are followed by a buffer which limits the amplitude, it can be assumed that the SSB noise is dominated by PM noise.

From (3.31), the SSB can be calculated

$$\begin{aligned} V(t) &= A \cdot \sin(\omega_0 t + \phi(t)) = A\sin(\omega_0 t)\cos(\phi(t)) + A\cos(\omega_0 t)\sin(\phi(t)) \\ &\approx A\sin(\omega_0 t) + A\cos(\omega_0 t)\phi(t) \end{aligned}$$

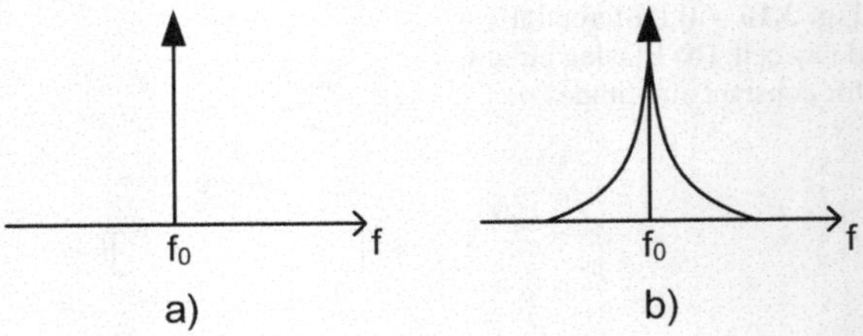

Fig. 3.17 (**a**) Voltage spectrum of an ideal oscillator (**b**) Voltage spectrum of a noisy oscillator

since $\phi(t)$ is assumed to be small, especially at moderate offset frequencies. The SSB noise is then

$$\mathcal{L}(\Delta f) = \left(\frac{|\mathcal{F}\{A\phi(t)\cos(\omega_0 t)\}|^2}{A^2/2}\right)_{f=f_0+\Delta f} = \frac{\frac{A^2}{4}S_\phi(\Delta f)}{\frac{A^2}{2}}$$
$$= \frac{S_\phi(\Delta f)}{2}$$

This relation is important since most calculations and simulations are easier done in the voltage domain from which the phase noise can be calculated.

For example, the phase noise power in a frequency band between f_1 and f_2 is

$$2\int_{f_1}^{f_2} S_\phi(f)df \tag{3.34}$$

while the integrated SSB power is

$$\frac{2\int_{f_1}^{f_2}\frac{A^2}{4}S_\phi(f)df}{\frac{A^2}{2}} = 2\int_{f_1}^{f_2}\frac{S_\phi(f)}{2}df \tag{3.35}$$

Note that, when a signal has random noise which is equally spread among amplitude noise and phase noise, only half of its noise power is phase noise. Furthermore, this relation only holds for small phase fluctuations. At low offset frequencies, the phase noise can be significantly high in free running systems. Though the phase Power Spectral Density (PSD) energy can be unbound, the voltage PSD cannot. In this case, the approximation $\sin(\phi(t)) \approx \phi(t)$ does not hold anymore and is expressed in the frequency domain through a Bessel function of the first kind.

When it comes to digital circuits and timing circuits, designers prefer time-domain measures like jitter as they can be more easily used in performance calculations. Though many definitions of jitter occur, the relation with phase noise in the frequency domain should be carefully analyzed.

If the system has bounded phase energy, then the rms jitter can be expressed

$$\sigma_{\tau_j}^2 = E\{\phi(t)^2\} = \frac{2}{\omega_0^2}\int_0^{\infty} S_\phi(f)df = \frac{4}{\omega_0^2}\int_0^{\infty} \mathcal{L}(f)df \tag{3.36}$$

However, a free running oscillator typically has a phase noise PSD proportional to $1/\Delta f^2$ which is not bounded. Indeed, if the phase is observed for a long time, its uncertainty will grow with time. To express free running systems in the time domain, K-cycle jitter or δ-jitter is commonly used which is the timing uncertainty after K cycles or a time δ .

$$\sigma_{\tau_j}(\delta)^2 = \frac{1}{\omega_0^2}E\{(\phi(t+\delta)-\phi(t))^2\} = \frac{2}{\omega_0^2}(R_{\phi\phi}(0)-R_{\phi\phi}(\delta)) \tag{3.37}$$

in which $R_{\phi\phi}(\delta)$ is the autocorrelation of $\phi(t)$ defined as

$$R_{\phi\phi}(\delta) = \int_{-\infty}^{\infty} S_\phi(f)e^{2\pi jf\delta}df \tag{3.38}$$

The jitter then reduces to

$$\sigma_{\tau_J}(\delta)^2 = \frac{8}{\omega_0^2}\int_0^{\infty} S_\phi(f)\sin^2(\pi f\delta)df \tag{3.39}$$

This shows that noise frequencies n/δ do not contribute due to their periodic behavior [106]. The function $\sin^2(\pi f\delta)$ is shown in Fig. 3.18. A special case is when $\delta = 1/f_0$. This is cycle-to-cycle jitter and is $\frac{1}{\omega_0^2}E\{(\phi(+1/f_0)-\phi(t)^2\}$

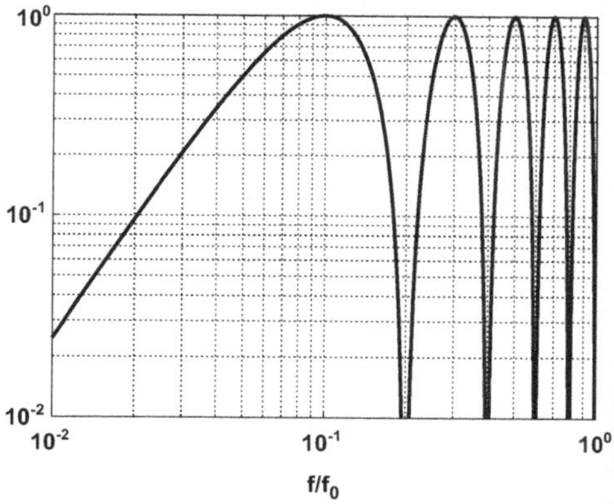

Fig. 3.18 Frequency plot of $\sin^2(\pi f\delta)$ for $\delta = 5/f_0$

$$\sigma_{cc}^2 = \frac{8}{\omega_0^2} \int_0^\infty S_\phi(f) \sin^2(\pi f / f_0) df \tag{3.40}$$

These equations transfer from phase noise calculations to the time domain.

3.4.2 *Phase Noise in LC-Tank Oscillators*

Mainly two different methods exist to calculate the phase noise of a harmonic oscillator. First is a linear approach which calculates the voltage noise across the tank and uses the assumption that this relates directly to the phase noise. This approach works well and gives good insight to the trade-offs in the circuit. However, it does not explain up conversion of low frequency noise. A second approach is a time-variant calculation which can take into account these effects.

3.4.2.1 Linear Calculation

In this approach, the impedance, close to the resonance frequency is used to calculate the SSB noise in an LC-tank oscillator [107]. The parallel impedance of an LC tank is

$$Z(j\omega) = \frac{sL}{s^2LC + s\frac{L}{R_p} + 1} \tag{3.41}$$

To calculate the voltage noise around ω_0, $|Z(j\omega)|^2$ can be approximated as

$$|Z(j(\omega_0 + \Delta\omega))|^2 \approx \frac{1}{4(\omega_0 C)^2} \left(\frac{\omega_0}{\Delta\omega}\right)^2 = \frac{(\omega_0 L)^2}{4} \left(\frac{\omega_0}{\Delta\omega}\right)^2 \tag{3.42}$$

The noise from an equivalent parallel loss R_{p-tot} is

$$\begin{aligned} dV_n^2(\Delta\omega) &= |Z(j(\omega_0 + \Delta\omega))|^2 \cdot \frac{4kT}{R_{p-tot}} \cdot df \\ &= kT \frac{1}{R_{p-tot}(\omega_0 C)^2} \left(\frac{\omega_0}{\Delta\omega}\right)^2 \cdot df \\ &= kT \frac{(\omega_0 L)^2}{R_{p-tot}} \left(\frac{\omega_0}{\Delta\omega}\right)^2 \cdot df \\ &= kT \frac{R_{p-tot}}{Q^2} \left(\frac{\omega_0}{\Delta\omega}\right)^2 \cdot df \end{aligned}$$

This equation can be rewritten to a series resistance R_{ser} instead of R_{p-tot} such that

$$dV_n^2(\Delta\omega) = kTR_{ser}\left(\frac{\omega_0}{\Delta\omega}\right)^2 \cdot df \tag{3.43}$$

in which R_{ser} is the equivalent tank series resistance

$$R_{ser} = R_L + R_c + \frac{1}{R_{parallel}(\omega_0 C)^2} \tag{3.44}$$

and in which R_L is the series resistance of the inductor, R_c is the series resistance of the capacitor, and $R_{parallel}$ is a parallel resistance that may be parallel to the tank.

Moreover, the active component G_m introduces noise which can be modeled as

$$di_n^2 = 4kTG_mF \cdot df \tag{3.45}$$

with F the noise factor of the amplifier. For a stable oscillation, $G_m = 1/R_{p-tot}$, though some safety margin α is introduced to ensure the oscillation starts.

Its total noise reduces to

$$di_n^2 = \frac{4kT\alpha F}{R_{p-tot}} \cdot df = \frac{4kTA}{R_{p-tot}} \cdot df \tag{3.46}$$

This noise source can be added to the total equivalent noise resistance

$$dV_n^2(\Delta\omega) = kTR_{ser}(1+A)\left(\frac{\omega_0}{\Delta\omega}\right)^2 \cdot df \tag{3.47}$$

dV_n^2 can be used to calculate the SSB noise

$$\mathcal{L}(\Delta\omega) = \frac{1}{2}\frac{dVn^2(\Delta\omega)}{P_{sig}} = \frac{1}{2}\frac{kTR_{ser}(1+A)}{V_{amp}^2/2}\left(\frac{\omega_0}{\Delta\omega}\right)^2 \tag{3.48}$$

which can also be rewritten to

$$\mathcal{L}(\Delta\omega) = \frac{2kTR_{p-tot}(1+A)}{P_{sig}}\left(\frac{\omega_0}{2Q^2\Delta\omega}\right)^2 \tag{3.49}$$

which is more known as Leeson's equation [105] in the $1/f^2$ region. The factor $1/2$ is added since only half of the noise power will contribute to phase noise.

From these equations, a few design choices can be derived. First is the amplitude of the signal. It is clear that this should be maximized to reduce the phase noise. Therefore, the bias current should be large enough to maximize the swing of the circuit. When the amplitude is limited by the bias current, the oscillator operates in the current limited region. When the current is further increased, the amplitude will be limited through saturation. This is the voltage limited region. In the voltage limited region, excess current does not improve the amplitude but does increase the noise. Therefore, the optimal bias point for the oscillator is when it operates at the edge of current limited mode.

At frequencies <5 GHz, most of the losses in the tank are determined by the inductor such that its series resistance is the main contributor to the phase noise (together with the noise from G_m which is coupled to R_L). Two design methods can be differentiated here. First is a design with a given power consumption, second is a design with a given phase noise.

In the first approach, the power consumption and bias current is fixed. The oscillation amplitude is proportional to $Q^2 R_L I_{bias}$ such that the SSB phase noise is

$$\mathcal{L}(\Delta\omega) \sim \frac{kTR_L(1+A)}{(Q^2 R_L I_{bias})^2}\left(\frac{\omega_0}{\Delta\omega}\right)^2 \tag{3.50}$$

which shows that an inductor should be chosen such that the amplitude is maximal, given a geometry with the best Q-factor. A smaller inductance reduces the voltage noise as well as the amplitude (squared). A higher inductance increases the noise without improving the swing.

In the case where the power consumption can be chosen, the maximal R_{eff} can be calculated from (3.48). The bias current is then chosen to maximize the swing.

When designing for best noise-to-power ratio, a tank should be chosen with the highest Q since this tank has the lowest noise for a given swing [108].

3.4.2.2 Impulse Sensitive Function

An approach to calculate the phase noise was introduced in [109–111] and relies on a time-variant method. When noise is injected in a tank, it does not always contribute to phase noise. When the signal is maximal, current which is injected in the tank contributes to AM noise but not to PM noise while, during zero-crossing, the noise contributes primarily to PM noise.

The idea is that an impulse charge, injected to the tank, creates a phase step and the amount of phase deviation depends on the moment when it is injected.

For this, a time-variant impulse response is proposed:

$$h_\phi(t,\tau) = \frac{\Gamma(\omega_0\tau)}{q_{max}}u(t-\tau) \tag{3.51}$$

in which $\Gamma(\omega_0\tau)$ is the Impulse Sensitive Function (ISF), $q_{max} = C_{node}V_{max}$ and $u(t-\tau)$ the unit step function. The ISF is a periodic function which represents the sensitivity of an oscillator to an impulse which is dimensionless.

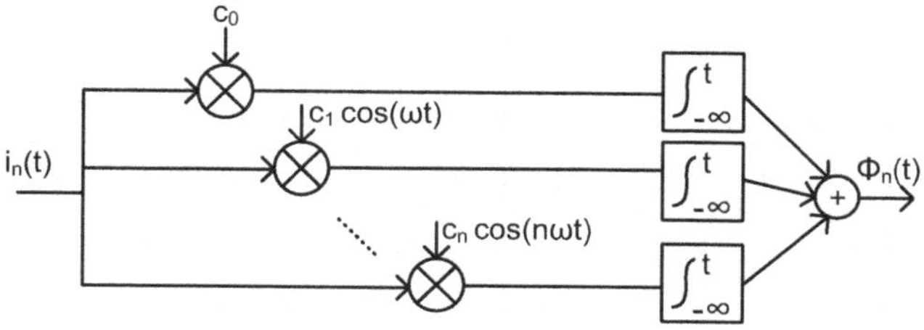

Fig. 3.19 Each branch represents a component of the ISF Fourier series which is mixed with $n\omega_0$ and filtered by an integrator

The total output phase noise due to a current noise source is

$$\phi(t) = \int_{-\infty}^{\infty} i_n(\tau)h_\phi(t,\tau)d\tau = \int_{-\infty}^{t} i_n(\tau)\frac{\Gamma(\omega_0\tau)}{q_{max}}d\tau \tag{3.52}$$

However, $\Gamma(\omega_0\tau)$ is a non-linear function of time. This causes noise mixing across higher harmonic frequencies in the spectrum.

The ISF can be expanded in a Fourier series such that the phase noise becomes

$$\phi(t) = \frac{1}{q_{max}}\left(c_0\int_{-\infty}^{t} i_n(\tau)dt + \sum_{n=1}^{\infty} c_n\int_{-\infty}^{t} i_n(\tau)\cos(n\omega_0\tau)d\tau\right) \tag{3.53}$$

in which c_n are the Fourier series components of the ISF. A graphical representation is shown in Fig. 3.19. Each branch is a mixer which mixes the input signal with $n\omega_0$ and then filters is with $1/s$ around DC. For example, coefficient c_1 downconverts $i_n(t)$ from frequencies around ω_0 and an integration filter is applied. Due to the integration from current to phase, white current noise converts to phase noise which is shaped with $1/f^2$. This explains why noise sources at $n\omega_0$ can still contribute to low frequency phase noise. Note however that the magnitude of c_n typically decreases with n.

A specific influence is exhorted by coefficient c_0 as it explains why low frequency noise can still be converted to phase noise which cannot be explained by the linear model. c_0 is the DC value of the ISF and is determined by the symmetry of the signal.

When the noise is white, the phase noise can be calculated in the frequency domain as

$$\mathcal{L}(\Delta\omega) = \frac{\frac{di_n^2}{df}\sum_{n=0}^{\infty} c_n^2}{4q_{max}^2\Delta\omega^2} = \frac{\frac{i_n^2}{df}\Gamma_{rms}^2}{2q_{max}^2\Delta\omega^2} \tag{3.54}$$

3.4.3 Phase Noise Spectrum of an Oscillator

From these equations, it becomes clear that white noise sources lead to a phase noise spectrum with $1/f^2$ dependency. Though also 1/f noise sources exist, especially at low frequencies. The ISF predicts that this noise is converted to phase noise (through c_0) and shaped with $1/f^2$ such that this is visible with a $1/f^3$ shape at the output of the oscillator as is shown in Fig. 3.20. Finally, a white noise region extends at high frequencies which is due to excess noise, similar to amplifiers.

3.5 Phase-Noise in PLLs

3.5.1 Noise Transfer Function

Noise in PLLs does not only originate from the oscillator. Besides the oscillator, the loop filter, charge-pump but also the reference clock contribute to the output phase noise. When operating in a closed loop, the noise, which is generated in the devices is shaped towards the output. This shaping depends on the loop dynamics of the PLL. The Noise Transfer Function (NTF) is a frequency domain transfer function which represents the relation between output phase noise and the noise sources in the circuit [112]. The dimension NTF can be [rad$^2/A^2$], [rad$^2/V^2$] or [rad^2/rad^2], depending on the type of the noise source.

Figure 3.21 shows the different noise sources (VCO, loop filter, charge-pump, and reference) in the PLL and their location in the loop [113].

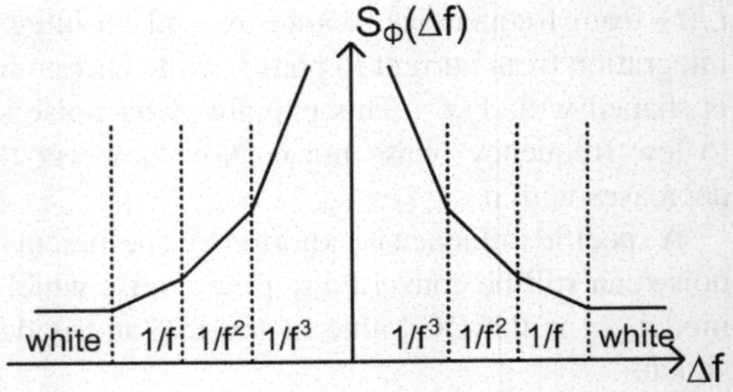

Fig. 3.20 Typical phase noise spectrum in an oscillator

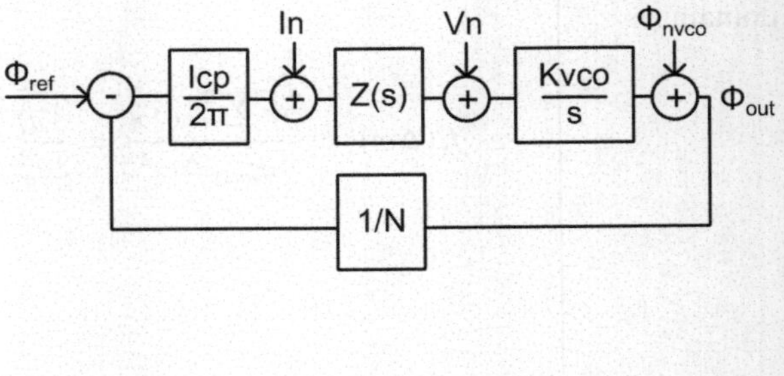

Fig. 3.21 Noise sources in a PLL

3.5.2 Reference Clock Phase Noise

Each reference clock will have some phase noise. This can be large or small depending on the application. In many clock synthesizers, a crystal reference is used to which the PLL is locked, these noise levels are typically very low and the integrated jitter can be as low as 100 fs. For such applications, typically the reference noise is neglected. In other applications, the reference clock may not be ideal and can be a major contributor to the noise.

The NTF of the reference phase to the output is simply calculated from the closed loop PLL transfer function from (3.5). The noise is low-pass filtered after the PLL bandwidth with −20 dB/dec for second order loops and −40 dB/dec for third order loops. Intuitively, the PLL will track slowly varying phase deviations but cannot track high frequency phase changes as the loop has no more gain at high frequencies. Furthermore, the phase noise is amplified by N^2 within the bandwidth since the PLL tracks time shifts which are higher phase differences at larger frequencies.

$$NTF(s) = \left| N \frac{\omega_n^2 (1 + \frac{2\zeta}{\omega_n} s)}{s^2 + 2\zeta\omega_n + \omega_n^2} \right|^2 \tag{3.55}$$

PLLs thus exhibit noise suppression from the input and many PLLs are used just to filter the noise of the reference clock to "clean-up" the phase noise. The smaller the PLL bandwidth, the more reference phase noise is suppressed. However, it will become clear that the NTF of the VCO is high-pass filtered leading to a trade-off between reference and VCO noise suppression.

3.5.3 Charge-Pump: Loop Filter

The noise of the loop filter is determined by the resistor used in the loop. The NTF from a voltage source, which is series connected to the resistor is

$$NTF(f) = \left| \frac{s K_{vco}}{s^2 + 2\zeta\omega_n + \omega_n^2} \right|^2 \tag{3.56}$$

which is band-pass filtered around $\omega = \omega_n(\sqrt{(2\zeta^2 - 1)^2 + 1} - (2\zeta^2 - 1))$.

If only a first order filter is assumed, the output phase noise is

$$S_\phi(s) = 4ktR \left| \frac{s K_{vco}}{s^2 + 2\zeta\omega_n + \omega_n^2} \right|^2 \tag{3.57}$$

At high offset frequencies, this reduces to

$$S_\phi(s) = \frac{4kTRK_{vco}^2}{4\pi^2\Delta\omega^2} \tag{3.58}$$

which originates from the voltage noise causing FM modulation on the VCO which can be seen as an additional noise component to the VCO.

To limit the output phase noise, it follows from (3.57) that the resistance can be reduced to enhance the phase noise. For constant dynamics, $1/RC$ and I_{cp}/C have to be constant. A reduction of the resistance implies an increase of the capacitance and charge-pump current. This is similar to $\sqrt{kT/C}$ noise.

The charge-pump's active devices also generate noise which can be seen as a current source parallel to the charge pump. The NTF can be calculated from (3.5):

$$NTF(s) = \left|\frac{2\pi}{I_{cp}}N\frac{\omega_n^2(1+\frac{2\zeta}{\omega_n}s)}{s^2+2\zeta\omega_n+\omega_n^2}\right|^2 \tag{3.59}$$

which has the same filter characteristics as the reference noise. Larger charge-pump currents reduce the phase noise but this also requires a larger loop capacitor. Note that, the charge-pump is only on for a short period of time $\approx T_{reset}$. This makes the noise contribution cyclostationary [114]. As a result, the noise power is reduced by T_{ref}/T_{reset} and the frequency domain noise is shaped with a convolution with the noise window.

The in-band phase noise can be approximated as

$$S_\phi(s) = 4kT\gamma G_m C\left|\frac{2\pi}{I_{cp}}N\right|^2 \tag{3.60}$$

in which C is the cyclostationary noise reduction factor due to the limited switching time and G_m the total equivalent noise transconductance of the current sources.

3.5.4 Oscillator

Oscillator phase noise is added directly at the output of the loop but low frequency noise is compensated by the feedback loop. The NTF of this noise source is

$$NTF(s) = \left|\frac{s^2}{s^2+2\zeta\omega_n+\omega_n^2}\right|^2 \tag{3.61}$$

which is a high-pass filter with a cutoff frequency at $\omega = \omega_n(\sqrt{(2\zeta^2-1)^2+1}-(2\zeta^2-1))$. Low frequency phase noise is filtered with 40 dB/dec while high frequency noise is passed to the output as was explained with the reference phase noise. High PLL bandwidths filter more noise originating in the VCO and suppress less reference noise while low PLL bandwidths filter more reference noise but less VCO noise.

3.5.5 Spurious Tones

Besides noise, spurious tones (tones with a single frequency) may occur in the phase noise spectrum. These tones originate from periodical signals in the PLL. For example, the reference spur which occurs due to the charge-pump switching is a periodic signal that creates tones in the spectrum. Furthermore, these signals are highly non-linear resulting in multiple harmonics.

Many signals will couple through the power supply or the substrate. For example, the divider has significant power consumption at the reference frequency since all registers will toggle simultaneously once per cycle (for synchronous dividers), while half of the registers toggle twice per cycle. This leads to significant spurs at f_{ref} and $2f_{ref}$. Another significant contributor to the noise coupled to the PLL are the input- and output stages which consume peaked currents during switching.

These effects can be reduced by proper VCO design which has good power supply rejection ratio (PSRR) and separated power supplies for the digital circuitry. Nevertheless, some noise will always be coupled through parasitic structures in the device. The same is true for other (digital) circuits when the PLL is used in a large SoC which can degrade its performance.

3.6 Performance Parameters

In general two figure-of-merits (FOMs) exist to compare the noise performance of VCOs [115]. The first FOM relates the phase-noise to power:

$$FOM_{PN} = -10\log(\mathcal{L}(\Delta\omega)) + 20\log\left(\frac{\omega_0}{\Delta\omega}\right) - 10\log(P_{1\,\mathrm{mW}}) \tag{3.62}$$

However, tuning range may also be included in this FOM:

$$FOM_{PN,tune} = FOM_{PM} + 10\log\left(\frac{TR}{V}\right) \tag{3.63}$$

in which TR/V is the tuning range per volt of the VCO.

The FOM for VCOs are usually calculated in the $1/f^2$ region. Since PLLs have a highly shaped output noise spectrum, it becomes difficult to define a location in the frequency domain to benchmark the circuit. Therefore, PLLs are typically benchmarked with a FOM:

$$FOM_{PLL} = 10\log\left(\frac{\sigma_{\tau_j}^2}{1s} P_{1\,\mathrm{mW}}\right) \quad (3.64)$$

in which σ_j is the rms jitter at the PLL output [115].

3.7 Conclusion

This chapter introduced the fundamentals of charge-pump PLLs and its non-ideal effects. The phase domain model and its loop dynamics were discussed with circuit implementations of the components in a PLL. These relations will be further used when the radiation effects on these circuits are discussed. In the second part, the phase noise and jitter in a PLL were discussed. A linear and time-variant model is used to calculate the phase noise in an oscillator in which the latter can explain up- and downconversion of the noise in an oscillator. This time-variant model will become useful when single-event effects are introduced to the oscillator. Finally, the different noise sources in a closed loop PLL are discussed with their effect on the output phase noise.

Chapter 4
Single Shot Time-to-Digital Converters

4.1 Introduction

This chapter discusses the system level design and physical implementation of a single-shot TDC. In Chap. 2, various topologies were discussed to implement a Time-to-Digital converter, each of them having its own advantages and disadvantages. Since the application scope of this work is high-energy physics and space, the most important constraints are radiation tolerance, environmental effects, production yield, and of course performance.

Compared to normal consumer electronics, in which the temperature range is 0–85 °C, these applications usually target military specifications from −55 °C up to 125 °C. This extended temperature range introduces stringent design constraints that may limit the performance. Usually, many identical circuits are used in the application, especially in high-energy physics detectors where 1.000–10.000 identical readout channels are deployed to measure the experimental interactions of the particles [116, 117]. The architectures should therefore be tolerant against process, voltage, and temperature variations and should meet the same performance in the various operating conditions with an acceptable yield.

The major performance indicator of a TDC is typically the timing resolution that can be achieved in the data converter. In modern high-energy physics applications, a resolution below 10 ps is a typical target. Additionally, the power consumption of the design is an important consideration. In particle detectors, the overall power consumption of the electronics is limited by the DC voltage regulators that distribute the power to the different channel modules, as well as heat generation by the electronics. While particle detectors do not suffer from a very strict power budget (limited though [118]), space applications have a very tight power budget [119]. Typically, satellites, probes, or launch vehicles are battery- or solar powered and only have limited power available which has to be shared among all system components. Furthermore, the heat dissipation is only through radiative heat emission [120].

J. Prinzie et al., *Radiation Hardened CMOS Integrated Circuits for Time-Based Signal Processing*, Analog Circuits and Signal Processing,
https://doi.org/10.1007/978-3-319-78616-2_4

Finally, ionizing radiation will play a significant role in the architectural design. Although there are many solutions to cover radiation effects at circuit or layout level, a system level design which is intrinsically less sensitive to radiation will benefit the design. TID effects may degrade the performance over time. Since accurate time measurements are required, self-calibration is necessary. Because the TID effects change the transistor parameters over time, an initial calibration (e.g., after production tests) is not sufficient. Single Event Effects furthermore degrade the performance at any point in time. In single-shot TDCs, the offline time due to an ionizing particle should be minimized. Often, there is only a single chance to do a measurement and the data will be lost when the circuit is in a reinitialization mode.

Taking into account these constraints has led to an architecture that will be discussed in this chapter. Among the various architectures in literature, the delay line or Voltage Controlled Delay Line (VCDL) based topology proved to be a good candidate for a radiation tolerant design.

4.2 TDC System Level Architecture

4.2.1 Self-Calibration Loop Implemented by a DLL

A TDC based on a VCDL can, in general, be split into two main blocks as is shown in Fig. 4.1. First is the timing generator which is composed of a VCDL that generates the different phases, equally spaced in time. These phases, or timing bins, are fed to the various channels of the TDC. Most of the performance of the TDC is determined in this timing generator block. The second block is the channel array. Each channel takes a snapshot of the timing generator's phases when it is triggered by the channel trigger. Based on this snapshot, a digital value can be decoded that represents the time difference between the channel trigger and the reference clock of the VCDL. The channel trigger is the start or stop channel in a 2-channel TDC. This approach facilitates the use of multiple channels that share a single timing generator [121]. In a more complex design, more than 64 channels can be employed. In a next step, a digital back-end processing block can calculate the time differences between the different channels [122].

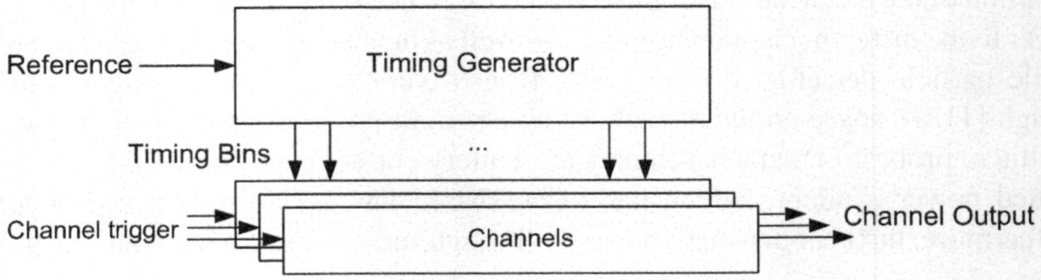

Fig. 4.1 Generic block diagram of a VCDL based, multichannel TDC

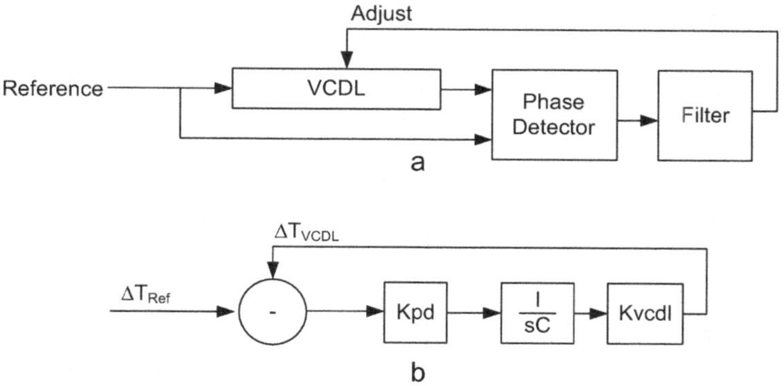

Fig. 4.2 Simple model of a delay locked loop

The phases inside the timing generator are generated from a VCDL that is controlled by a delay locked loop (DLL). A DLL is a control loop which ensures that the total delay of the VCDL is equal to the period of the reference clock. The DLL ensures that the resolution of the TDC is constant, even after irradiation damage. A typical DLL block diagram is shown in Fig. 4.2a [123]. A reference clock is injected in the delay line. A phase detector measures the phase difference between the output of the VCDL and the reference clock. If the system reaches a stable state, the nth clock cycle that propagates into the VCDL is in phase with the $(n+1)$th cycle of the reference clock. The loop filter, typically an integrator, averages the phase error and this signal is fed back to the VCDL. Since there is an integration in the loop, the error at the output of the phase detector converges to zero which means that the input clocks of the phase detector are aligned.

The input and output signals of the DLL are periodic clock signals. Essentially, the phase difference between those signals is interesting in the design. A VCDL can be modeled as a gain of K_{VCDL}, as is shown in Fig. 4.2b. This gain can be calculated through simulations and is the slope of the delay to voltage curve. Furthermore, the phase detector can be initially modeled as a subtraction. It is assumed that the phase detector generates a signal which is proportional to the difference between the VCDL delay and the period of the reference clock. The integration can be implemented with a charge-pump that sources or drains a current from a capacitor. The gain of this block is therefore I_{cp}/C.

The open loop transfer function of the DLL can be calculated as:

$$H_{OL}(s) = \frac{K_{PD} I_{cp} K_{VCDL}}{sC}. \tag{4.1}$$

The closed loop transfer function then simplifies to:

$$H_{CL}(s) = \frac{\Delta T_{VCDL}}{\Delta T_{ref}} = \frac{1}{1 + \frac{sC}{K_{PD} I_{cp} K_{VCDL}}} = \frac{1}{1 + s/\tau} \tag{4.2}$$

in which ΔT_{VCDL} is the delay of the VCDL and ΔT_{ref} is the period of the reference clock. First of all, this is a single pole system, which is always stable (compared to a PLL which is not). Secondly, it is interesting to look at the various parameters in the time constant of the loop.

$$\tau = \frac{K_{PD} I_{cp} K_{VCDL}}{C}. \tag{4.3}$$

By increasing the charge-pump current or decreasing the capacitor, the speed of the loop and also the acquisition time are reduced. This will be, as explained later in this section, a design constraint in DLLs. The speed can also be boosted by increasing the tuning range of the VCDL. As the tuning range is increased, so is the slope and K_{VCDL}. A fast startup is required for a fast re-acquisition after a high-energy particle hits the circuit and pulls the DLL out of lock. The faster the circuit re-acquires lock, the less the system will be off-line.

While many circuit implementations are possible for the VCDL and charge-pump, they all reduce to the same linear model as introduced before. However, this is not true for the phase detector. Many phase detectors were used before but overall, they can be divided into two groups. The first group contains the linear phase detectors. As is shown in Fig. 4.3a, the output signal of the phase detector is proportional to the input phase difference. The smaller the phase error, the smaller the correction signal. If every component in the DLL is assumed to be ideal, the transient startup signals die and there is no limit cycle in lock. A second group is the non-linear or bang-bang phase detectors (BBPDs), shown in Fig. 4.3b. The output is either +1 or −1. No matter the phase error, the output will be +1 or −1, also called early or late.

This leads to two drawbacks. Firstly, during startup, the correction current to the capacitor is always the same in a bang-bang phase detector. Secondly, since this correction current is always the same, there exists a limit cycle around a locked state

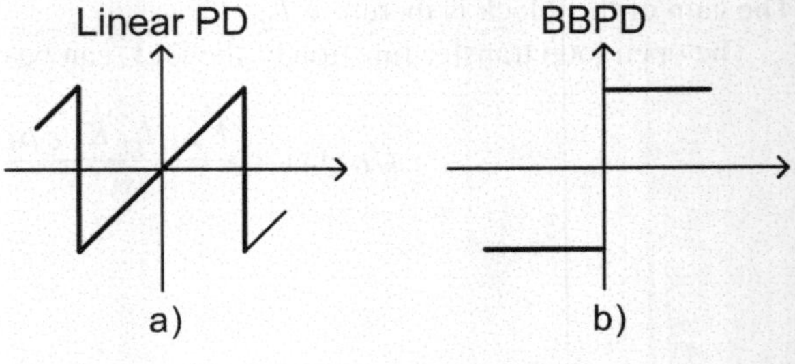

Fig. 4.3 (**a**) Linear (Phase Frequency Detector) phase detector. (**b**) Non-linear bang-bang phase detector

because there is always a positive or negative correction. The peak-to-peak jitter that is associated with this limit cycle is

$$J_{p2p} = \frac{I_{cp} T_{ref} K_{VCDL}}{C}. \tag{4.4}$$

The limit cycle of the BBPD introduces a serious trade-off between the control-voltage ripple and the (re)acquisition speed [124]. Since the current sources, driven by the BBPD, are continuously charging the loop capacitor, the ripple depends on the absolute value of this current (and capacitor). Low current operation results in a low control voltage ripple but also in a slow startup speed, whereas high currents improve acquisition speed but degenerate deterministic noise of the VCDL (control ripple). A trade-off has to be made between speed and deterministic noise. Doubling the startup speed would also double the voltage ripple in the limit cycle.

In contrast, a linear phase detector does not have this problem of limit cycle operation. During acquisition, a large correction of the phase detector/charge-pump quickly converges the loop. The net output current decreases near the locking point resulting in steady operation. However, linear phase detectors suffer from large static phase offsets resulting from the mismatch between the charge-pump currents. These offsets can increase up to several 10 ps. A comparison between the startup- and lock characteristics of BBPDs and linear phase detectors is shown in Fig. 4.4. It is clear that the startup dynamics of a linear phase detector are superior to the BBPD but that the in-lock phase offset of a BBPD is more interesting.

The proposed TDC uses a combination of both detectors to ensure a low static phase offset from the BBPD and the high startup speed from the linear phase detector. This eliminates the trade-off between startup speed and control voltage ripple of the DLL. In this design, it has been chosen to use a traditional phase-frequency detector as a linear phase detector in combination with a BBPD forming a hybrid phase detector.

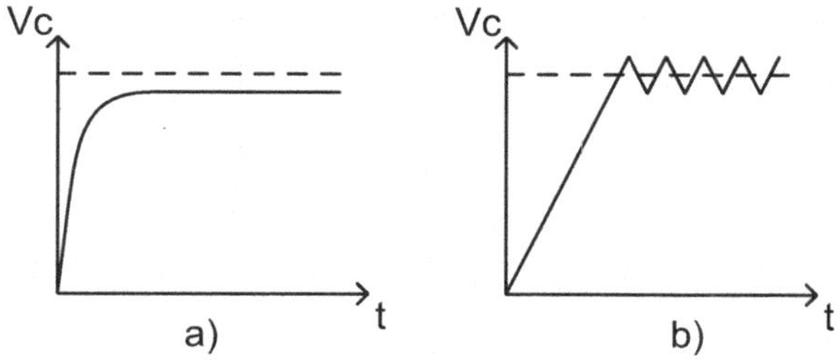

Fig. 4.4 (**a**) Linear phase detector. (**b**) Non-linear bang-bang phase detector

4.2.2 System Architecture with Double Phase Detector

The combination of a PFD with a BBPD requires some special attention. It has to be assured that the two phase detectors are not operating in a concurrent fashion. This means that one of the two detectors should be dominant at each point in time. During startup, the linear phase detector has control over the loop until the DLL is nearly locked. When it is close to lock, it is essential that the linear phase detector does not correct the loop anymore since it may continue to introduce static phase offsets.

A block diagram of the DLL architecture with a hybrid phase detector is shown in Fig. 4.5. Both phase detectors are combined and process the VCDL and reference signal independently. The PFD has a built-in deadzone and drives a high current charge-pump (50 μA) compared to the BBPD. In this way, during start-up, the PFD dominates the loop dynamics with a high speed. When the loop is approaching its locking point, the PFD enters its built-in deadzone. The deadzone ensures an automatic shutdown and reactivation of the PFD without any complex detection scheme. In this design a 20 ps deadzone window has been chosen. After the PFD shuts down, the BBPD only has to acquire lock within this narrow timing range. Now, only the BBPD is in control of the loop since the PFD is no longer correcting the phase errors. The low current (2 μA) charge-pump, driven by the BBPD ensures a low ripple in the limit cycle. In the case of a loss of lock due to an ionizing particle, the PFD leaves its deadzone window and is reactivated automatically. The loop will then be controlled again by the PFD and will quickly recover until the PFD enters its deadzone window again.

The deadzone of the PFD has been accurately designed using a PFD without deadzone using dedicated deadzone logic in both up and down signals as shown in Fig. 4.6. These up and down signals are compared using a NAND gate that will generate a pulse with a pulse-width proportional to the phase error. However, for small phase errors (within the targeted deadzone window), the NAND gate will not be able to reach the switching point of the inverter and no pulse will be generated at the output. The deadzone window can be controlled with the slew rate of the NAND

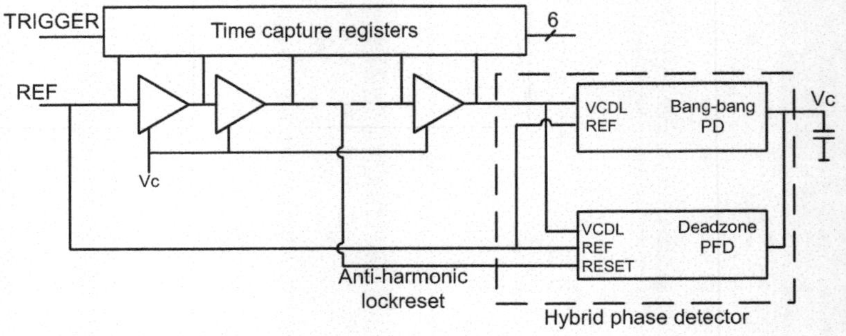

Fig. 4.5 Proposed DLL architecture with hybrid phase detector

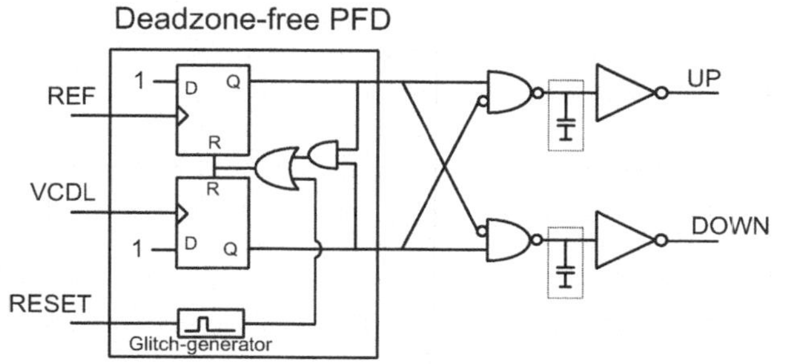

Fig. 4.6 Detailed PFD with controlled deadzone and cyclic reset

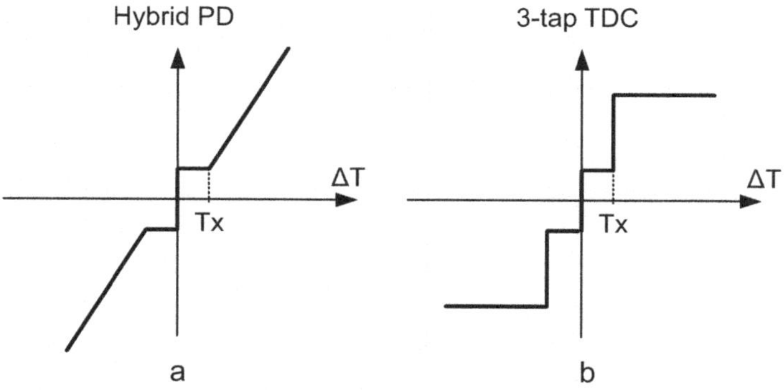

Fig. 4.7 Transfer functions of the phase detector with charge-pump for (**a**) the proposed hybrid detector, (**b**) 3-tap TDC phase detector

gate and depends on the drive current of the gate and its loading capacitance. If the load capacitor or the NAND gate current is made programmable, a tunable deadzone window can be implemented.

Another solution to boost the startup speed of the DLL is to use a 3-tap TDC as a phase detector which provides signals "early/late" and "very early/very late" [125]. The window of 20 ps that was built in the PFD can be used here as well by designing the 3-tap TDC with a bin size of 20 ps. When the state of the phase detector is very early or very late, a high current charge-pump is activated, in the other case, a low current charge-pump is activated. A comparison between both detectors is shown with their corresponding transfer functions in Fig. 4.7. The proposed hybrid phase detector is linear in the region $|\Delta T| > T_x$ and has a discrete functionality in the region $|\Delta T| < T_x$. The 3-tap TDC has identical functionality in locked state ($|\Delta T| < T_x$) but the dynamics differ outside $|\Delta T| > T_x$. There, the current is also saturated, but larger. The acquisition time depends on the initial phase error since the current (which may be large) is still constant. This phenomenon is not the case in a linear PD where the loop nearly reaches a converged state after 5τ, in which τ is the time constant of the DLL loop.

Some special attention is required on the PFD. This is a block that originates from PLL designs in which it has proven its use for many years. Essentially, a PFD is a state machine that provides up and down signals with a pulse width that is proportional to the time difference between the two inputs. Due to the state machine, it has a memory-like behavior and a cycle slip may be required to bring the PFD to a correct state. This depends on the initial state of the PFD before it starts its phase detection. These cycle slips are possible in PLLs but not in DLLs since the output clock is always a delayed version of the reference clock. If the PFD would require a cycle slip, the DLL will converge to a dead lock state, either the control voltage will saturate at the ground or the supply rails.

To prevent this, an initial reset of the PFD may be sufficient in non-ionizing radiation environments. However, concerning single-event upsets, the state of the PFD may be upset due to an ionizing particle at any point in time. Therefore, the reset is required to be issued periodically, in this design, every clock cycle. Figure 4.5 shows the reset of the PFD which is an intermediate clock phase at the center of the VCDL. In this way, it is always ensured that the reset will rise before the output of the VCDL rises.

Two extreme timing situations may occur [126]. The delay of the VCDL is either minimal (T_{min}) or maximal (T_{max}). If the delay is minimal, the PFD is reset after $T_{min}/2$ and the VCDL output signal rises at T_{min}. There is no constraint on the minimal delay of the VCDL to prevent a false lock. If the delay is maximal, the PFD is reset after $T_{max}/2$. This should happen before the next clock cycle at T_{ref}. Therefore the maximal delay of the VCDL should be limited to $2T_{ref}$. An extra margin has to be taken into account for the reset delay circuitry of the PFD and local clock delays.

4.2.3 TDC Timing Generator Linearity

Ideally, each element in the VCDL has an identical delay. However, in reality a random variation occurs between each tap in the VCDL due to random variations of processing. This leads to INL and DNL errors and contributes to the non-linearity of the TDC. Each cell has a delay $T_d[n]$ which is composed of a nominal delay T_{cell} with a random variation ϵ. The delay of the VCDL at the nth tap is

$$T_{VCDL}[n] = \sum_{k=1}^{n} T_d[k] = \sum_{k=1}^{n} T_{cell} + \epsilon[k] = nT_{cell} + \sum_{k=1}^{n} \epsilon[k]. \tag{4.5}$$

Recall from Chap. 2 that a delay line based TDC has an uncertainty at the nth stage with a standard deviation of $\sqrt{n}\,\sigma(\epsilon)$.

The result is that the uncertainty is statistically maximal at the end of the delay line (in an open loop design). However, if the VCDL is controlled by a closed loop DLL, the end of the VCDL is locked to a known time (which is the period of the reference clock). Therefore, in a closed loop, the DLL will correct for the total accumulated variation in the delay cells. If a delay line with N taps is used, the total delay equals

$$T_{VCDL}[N] = NT_{cell} + \sum_{k=1}^{N} \epsilon[k] = T_{ref}. \quad (4.6)$$

The delay of each cell is then adjusted with a value

$$T_{cellcorr} = \sum_{k=1}^{N} \epsilon[k]/N. \quad (4.7)$$

The uncertainty after n stages reduces to

$$\epsilon_{DLL}[n] = \sum_{k=1}^{n} \epsilon[k] - \frac{n}{N} \sum_{k=1}^{N} \epsilon[k] = \left(1 - \frac{n}{N}\right) \sum_{k=1}^{n} \epsilon[k] - \frac{n}{N} \sum_{k=n+1}^{N} \epsilon[k] \quad (4.8)$$

The standard deviation at the nth stage is

$$\sigma_{\epsilon_{DLL}}(n) = \sigma_{\epsilon} \sqrt{\frac{n(N-n)}{N}} \quad (4.9)$$

By deriving this function and equating it to zero, the maximum variation can be found to be $\sigma(\epsilon)\sqrt{N}/2$ in the middle of the delay line ($N/2$). A comparison between an open- and closed loop delay line is shown in Fig. 4.8. The reduction in INL error and predetermined bin size makes the DLL a superior circuit in systems with a large amount of channels where pre-calibration of the INL is difficult and delays can be degraded due to TID effects. The variations are then compensated by the DLL. The only disadvantage of using the DLL is the required startup time and the additional circuitry which can be sensitive to ionizing radiation for which this improved loop is proposed.

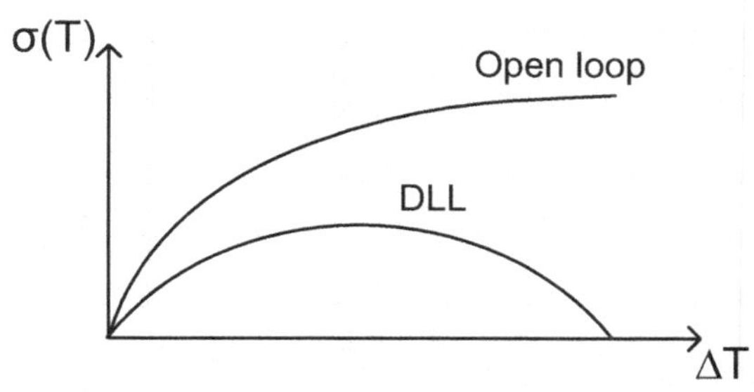

Fig. 4.8 Random delay variations in an open- and closed loop delay-line based TDC

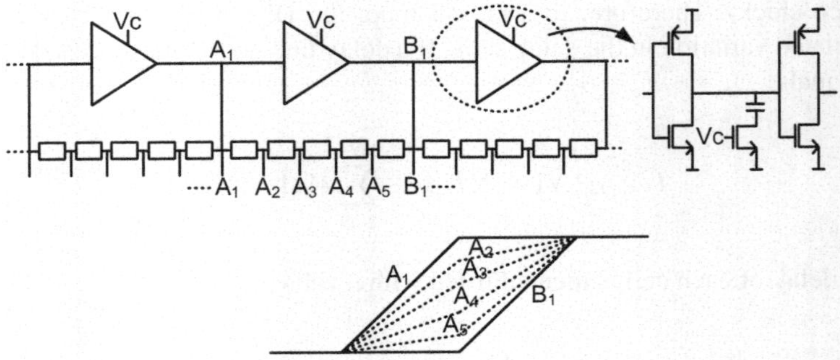

Fig. 4.9 Delay cell resistive interpolation

The linearity of the TDC is mainly defined by the variations of the delay cells in the VCDL. A resistive interpolation scheme is used in this design to achieve sub-gate delay resolution as shown in Fig. 4.9. Each delay of the DLL is interpolated by 5. The resistors' taps generate a weighted value of the input signal and output signal of the delay cell. The resistors are implemented with poly tracks and are scaled to ensure that the RC delays between all phases are equal [68]. In this chip, a single-ended delay cell architecture is implemented. Noninverting buffers are required because of the resistive interpolation. This will result in a speed decrease by a factor two, compared to inverting stages or differential delay cells. However, for the same power consumption, a single-ended cell architecture is less noisy if proper power supply decoupling is done. To improve the speed of the noninverting stages, only one inverter per pair is speed-limited, only for rising edges. While this results in an asymmetric duty cycle, the DLL has no issues with this asymmetry because only rising edges are processed. This results in a speed penalty of less than two.

4.2.4 Channeling and Basic Readout Interfaces

The DLL only acts as a timing generator that creates N-equally spaced phases of the reference clock and most of the critical blocks are located in this macro. Nevertheless, a channel has to be created that samples the VCDL followed by a digital block which decodes the VCDL snapshot to a binary representation. Furthermore, the dynamic range is only one clock cycle of the reference clock period (T_{ref}) but this can be extended with a counter that runs at the reference frequency.

The implementation of a TDC with a single channel is shown in Fig. 4.10. The channel takes a snapshot of the VCDL when the hit signal rises and this data is stored in the registers of the channel. These registers are further labeled as L0 (Level 0) registers of the TDC. With a single channel, the delay can be measured between the hit signal and the reference clock. However, usually the time difference between a start and stop signal has to be measured. For this approach, a DLL based TDC

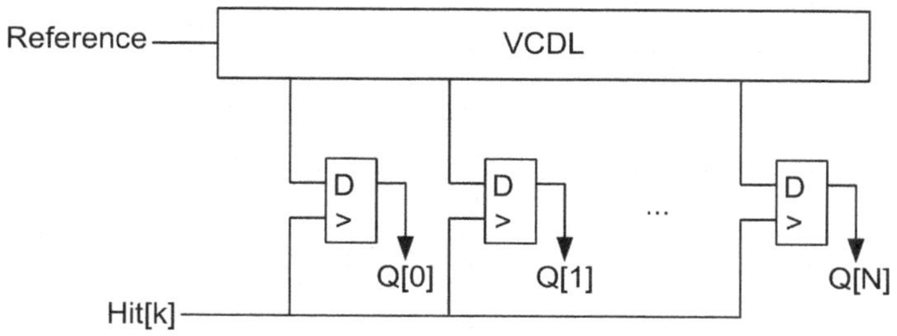

Fig. 4.10 Implementation of a single channel TDC

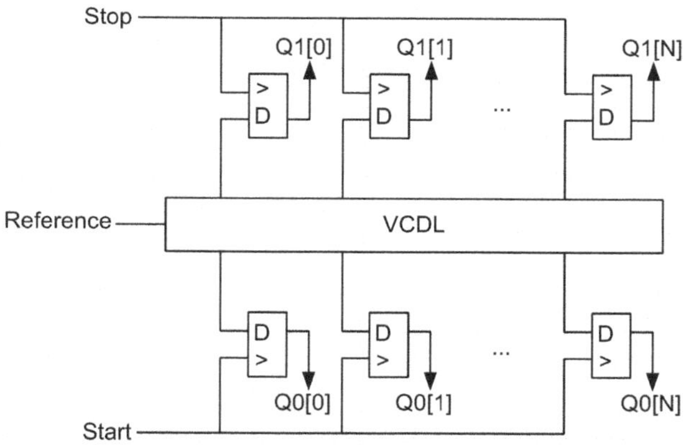

Fig. 4.11 Implementation of a start/stop channel TDC

requires 2 channels. One channel for start and stop, respectively. The time difference can then be calculated digitally. This approach is shown in Fig. 4.11.

The decoding of the VCDL snapshot can be fully done in a digital back-end running at slower frequency or can be combinational. The basic mechanism for the decoding is the conversion of a pseudo thermometric code to binary. In this TDC, the decoding is done by finding a "100" sequence in the sampled output data. This strategy enables the rejection of single bubbles in the VCDL snapshot. A bubble may occur due to metastable flip-flops or single event upsets in the flip-flops. This detection is done in the first layer of the logic as is shown in Fig. 4.12. The first layer will provide a positive output at the location where the "100" transition occurred. This output drives a lookup table which converts the location to a binary value. For safety, a priority encoding is inserted which prevents a false output value if occasionally, multiple outputs of the "100" detector layer are high. This may occur due to single-event upsets in one of the flip-flops.

This decoding can also be fully synthesized which reduces the design effort but usually also reduces the speed. A drawback of a synthesized block is the signal connection complexity from the analog macro to the digital macro. N signals are

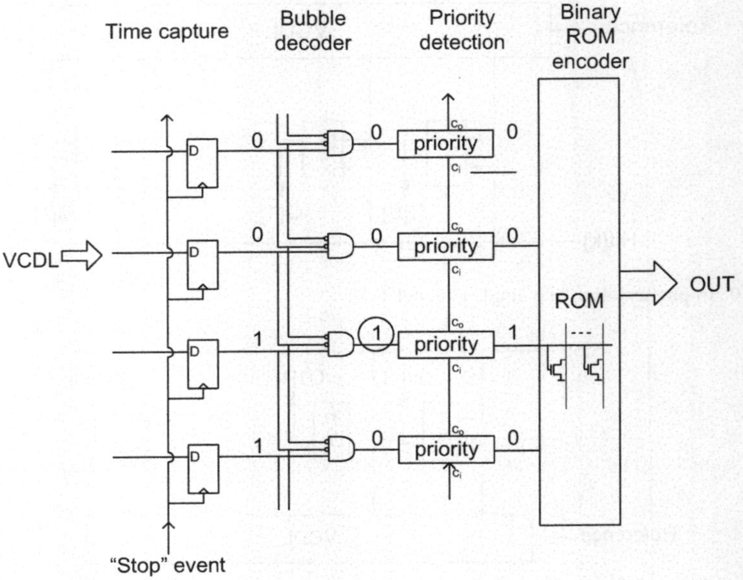

Fig. 4.12 Full custom decoder

required from the flip-flops to the digital block per channel. This may become impractically for TDCs with 64 or more channels. In that case, the decoding should be done locally, close to the flip-flops which sample the delay line.

To connect the flip-flop which sample on the stop signal to a digital back-end core, a first layer of synchronization is required to synchronize the registers to the clock domain of the digital block. This synchronization layer (L1 registers) provides the sampled VCDL data (L0 registers) to the decoder which searches for a "100" transition in the L1 data as is shown in Fig. 4.13a. Usually this block does not run at a very high frequency compared to the reference clock of the DLL. Though, if a maximum performance is required for a high throughput, the sampling rate of the data converter can be as high as the reference clock. At frequencies of 1 GHz or above, advanced decoding structures may be required to meet the timing of the design by splitting the decoding of N bits in multiple parallel blocks and pipeline the conversion. In this work, the conversion rate was limited to only 150 MSa/s.

The dynamic range of the TDC can be extended with a counter which runs at the reference clock since the VCDL can only represent the timing information that is limited within 1 cycle. To improve the linearity and noise of the DLL, a high speed reference clock is required [122]. High clock speeds place serious constraints on the digital counters, especially if a TMR structure is required which limits the speed to 1 GHz or slightly above in 65 or 40 nm CMOS technologies. The counter's value is sampled in a similar way as the VCDL is sampled when a hit signal arrives. However, the timing information in the counter and VCDL is encoded in

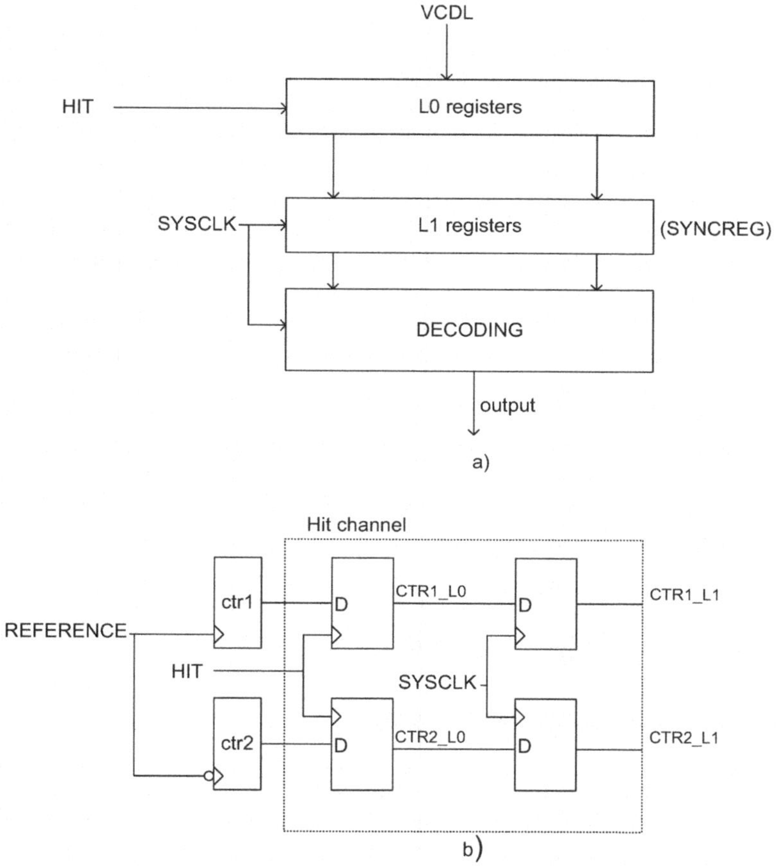

Fig. 4.13 (**a**) Digital decoder of a channel with L1 synchronization registers. (**b**) Dynamic range extension counters running at both clock edges

a different way. The VCDL has a pseudo thermometric coding. Since the hit signal is asynchronous, it may introduce metastability in flip-flops where setup or hold times are violated. In the pseudo thermometric code this may result in 1 LSB error when the hit signal arrives at the edge of a bin. The counters typically encode the count value in a binary fashion. Furthermore, the counters also represent the MSB information of the timed signals. Any errors in this data will result in significant malfunctions. When a counter changes its value after the rising edge of the clock, the output bits are unstable for some period of time. If the hit signal samples the counter within this time frame, the result will be incorrect.

The dynamic range extension that uses two counters is shown in Fig. 4.13b. One counter counts at the rising edge of the high speed clock, which also drives the DLL, while the other counter counts at the falling edge of the clock. Upon arrival of a hit signal, both counters are sampled into an L0 register. The idea of this approach is that at least one counter will be stable at any point in time. Therefore, when the hit signal arrives, one of the counters will provide a stable value. In the first part of the clock cycle, CTR1 will not be stable. It can be assumed that the counter is unstable from the launching time of the clock until the propagation delay of the flip-flops in the counter. The delay of any clock-tree is not included here since the same clock delay is assumed in the DLL and hit tree to balance all clock paths. The counter should be stable within half the clock cycle of the reference clock of the DLL. Since CTR2 toggles at the falling edge, it will be unstable in the other half of the clock cycle. When the hit signal has sampled both counters, either CTR1 or CTR2 has to be chosen. This decision can be made with the MSB of the decoded values from the decoder from the VCDL. The decoded signal from the VCDL represents the time difference between the HIT signal and the reference clock. The MSB can therefore identify if the HIT occurred in the first, of second half of the period. Finally, the correct counter value can be concatenated from the L1 synchronization registers with the decoder output to represent the timing information in a binary format. The necessary timing constraints will ensure that the counters are fast enough to be stable within one half of the clock cycle.

4.3 Low Offset Bang-Bang Phase Detector

4.3.1 Origin of Static Phase Offsets

Static offsets usually originate in phase-detection circuits or mismatch between delays. The bang-bang phase detector in the DLL can result in absolute gain errors of the TDC. Circuit calibrations or back-end signal processing can take this offset into account if careful characterization is done. Usually in applications with many timing channels, this calibration has to be done automatically. In many cases where the number of channels in the system increases up to 1000 or more, this calibration has to be done on-chip to reduce the system overhead and complexity.

The static phase offset, resulting from the bang-bang phase detector is shown in Fig. 4.14. Ideally, the crossing point where CK2 is detected early or late happens

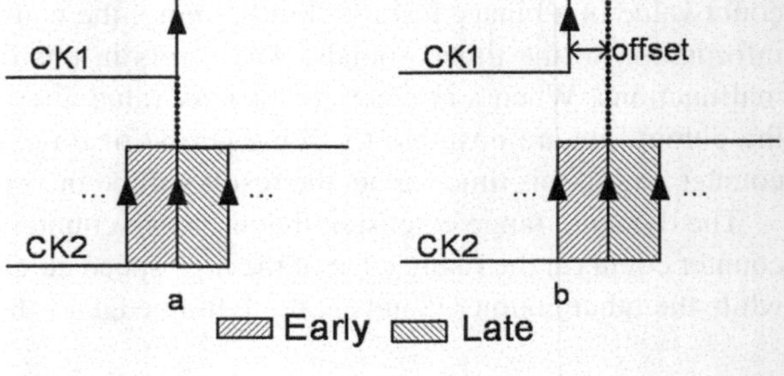

Fig. 4.14 (**a**) Ideal phase detector without static-phase offset. (**b**) Phase detector with static phase offset

at the CK1 event. However, because this detector is implemented using flip-flops or a sense amplifier, a delay mismatch or parasitic delays can result in an offset which can go up to several picoseconds. In time-based systems with resolutions in the picoseconds range, this can be a problem. Typically, large devices are used to ensure low offsets due to local mismatch but this comes at the cost of a high power consumption. Either the entire structure has to be scaled or extra buffers have to be added to drive the phase detector. Again these buffers suffer from delay mismatch.

The proposed phase detector uses a double state architecture that can either be in a calibration state or phase measurement state. The principle is based on well-known voltage domain offset cancellation techniques which are widely used in switched-capacitor amplifiers. In such circuits, the amplifier first measures its own offset and then reuses this information to amplify the signal in the second stage without offset [127] or uses a servo loop to compensate for this error [128]. In the proposed bang-bang phase detector, this principle is for the first time implemented in the time domain.

4.3.2 Removing Static Phase Offsets

Figure 4.15a shows the architecture of the self-calibrated phase detector. REF and VCDL are the reference and delayed clock signals of the DLL. The two-state operation of the detector is controlled by the signal "S". When $S = 1$, the circuit is in calibration mode. In this state, the input multiplexers both select the same clock signal CAL, which is a regular clock signal and can be a buffered version of REF. The calibration signal is injected in both paths of the phase detector. Ideally the delay mismatch at points A and B is zero. Depending on the offset of the core flip-flop, a decision is made. For positive offsets on the flip-flop, an identical clock on both inputs results in an "Early" decision, for negative offsets, a "Late" decision is made. In this way, the phase detector coarsely measures its own offset. The decision made by the core flip-flop is then stored into the lower flip-flop when S changes. This flip-flop controls a charge-pump for an inner calibration loop which is either in an "up" or "down" state. The current is integrated on a capacitor to control the delay elements in both input paths of the phase detector through a type I control loop. The delay cells in the input path work in a complementary way and compensate for the offset in the core flip-flop, controlled by this feedback loop. The decision, which is made in the calibration state, gives a negative feedback to this delay cells. Thus, each calibration cycle, the delays in the input path are adjusted until the delay difference compensates the internal offset of the phase detector. Further calibration events result in a limit cycle around a zero offset state of the phase detector. When $S = 0$, the circuit is in regular phase measurement mode and makes a decision whether VCDL is early or late compared to REF. Both VCDL and REF signals reach the phase detector through the two multiplexers and are given an extra delay which is equal to the offset of the core flip-flop but opposite. The decision of this measurement is stored in the upper flip flop and drives the charge-

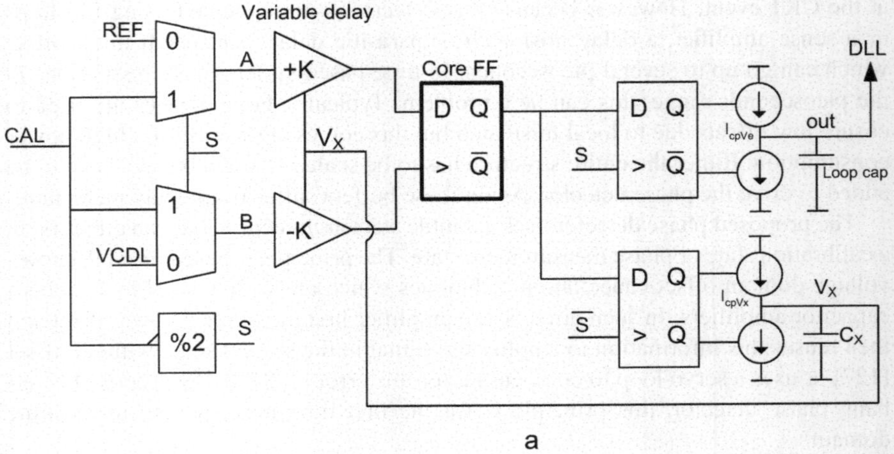

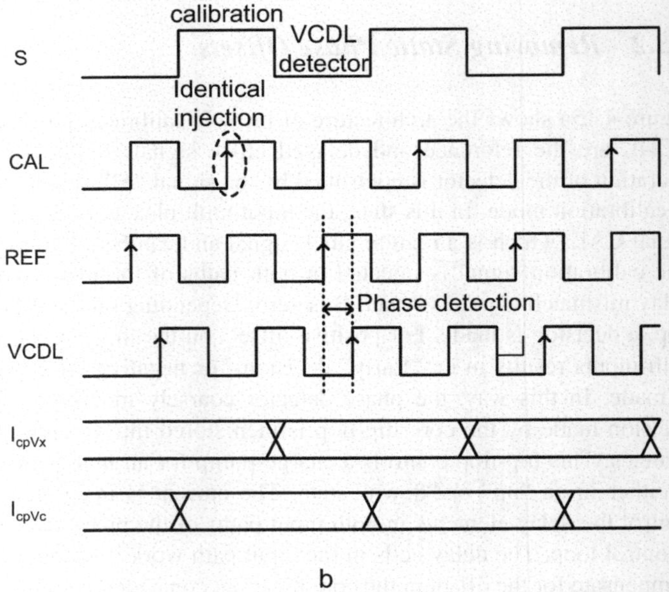

Fig. 4.15 (**a**) Self-calibrated phase detector implementation. (**b**) Signal waveforms of the phase detector in alternating states

pump and capacitor of the DLL loop. Since the signals REF and VCDL encounter a delay prior to phase detection, the offset that was generated by the flip-flop is now compensated. A signal waveform summary of the phase detector in both states is shown in Fig. 4.15b.

In this design it has been chosen to switch the state of the flip flop each cycle of the reference clock. Due to the limit cycle in the calibration state, deterministic periodic jitter occurs within the input of the phase detector. This jitter has a peak-to-peak value of

$$J_{\mathrm{p2p}} = \frac{2I_{\mathrm{cp}}T_{ref}K}{C_x} \tag{4.10}$$

in which I_{cp} is the calibration loop charge-pump's current, and K the voltage-to-delay gain of the delay cells in the input path. This jitter can be made arbitrarily small by using a large capacitor for the internal loop. However, since the internal loop has its own bandwidth, reducing this deterministic jitter also reduces the bandwidth of the inner calibration loop. A high loop bandwidth is preferred because low frequency noise (e.g., 1/f noise), generated in the phase detector, is suppressed within the loop bandwidth. However, for practical values of I_{cp} and C_x, this periodic cycle has the same order of magnitude as the input jitter of the clocks. The limit cycle is therefore randomized by the noise of the input path. In this design a current of 2 µA, 2 pF MOM-capacitor (C_x) and a K of 20 ps/V was chosen with a 2 GHz reference clock. This results in a calibration time of 200 ns. In this design, a MOM-capacitor has been used to prevent large leakage currents that can convert the type-I calibration loop into a non-ideal integrator.

The proposed DLL with inner phase detector calibration loop appears to be similar to a dual loop control system which is not unconditionally stable. An equivalent model of the two-state phase detector is shown in Fig. 4.16a. Fortunately, this topology does not suffer from the instability of the dual loop architecture. As long as both loops are individually stable, the entire DLL loop will be stable as well because there is only an inter loop connection from the calibration loop to the DLL loop and not vice versa. During startup, both loops will initialize but the calibration

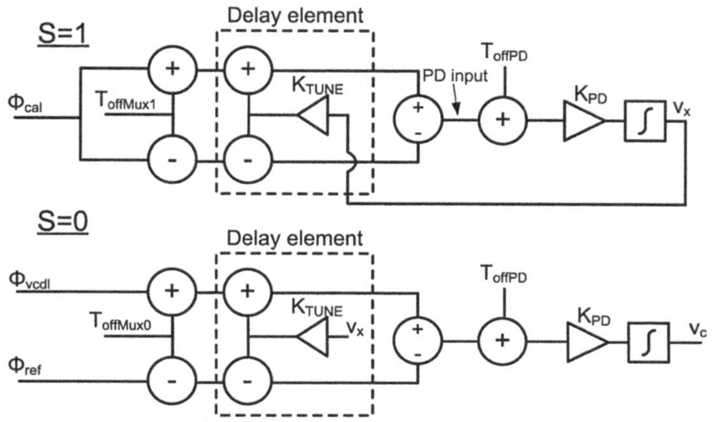

Fig. 4.16 Equivalent model for the calibration state and phase detector state

loop is not influenced by the DLL behavior, and they converge individually. The DLL however will encounter some startup transients since the delays at the input path of the phase detector are changing due to the calibration. However, once the calibration has converged, the DLL's transfer function is not altered because there is no connection from the DLL to the calibration loop and the calibration is transparent to the DLL operation. It is clear from this model that low frequency noise below the calibration loop bandwidth that originates within the delay elements or the phase detector is suppressed by the calibration closed loop transfer function.

4.3.3 Circuit Implementations

The mismatch of the proposed offset calibration technique is only limited by the mismatch in both multiplexers at the input. Ideally it has been assumed that no mismatch occurs between nodes A and B in Fig. 4.17. However, during calibration there will be a timing mismatch between the calibration signals at nodes A and B. The delay of the CAL signal to A and B is denoted as T_{a1} and T_{b1}, respectively. The mismatch between T_{a1} and T_{b1} will propagate inversely into the input paths of the phase detector since the loop compensates for it. This offset will remain inside the variable delay cells. A second offset will occur due to the mismatch between T_{a0} and T_{b0} during phase detection. The offset seen at the input nodes REF and VCDL is

$$T_{offset} = (T_{b1} - T_{a1}) + (T_{a0} - T_{b0}). \tag{4.11}$$

Since T_{a0} and T_{a1} are highly correlated, we can rewrite $T_{a0} = T_{a1} + \delta_a$ where δ_a represents the local delay mismatch within the multiplexer. The variation on the offset can be described as

$$\sigma_{T_{offset}} = \sigma_{\delta_a - \delta_b} = \sqrt{2}\sigma_\delta \tag{4.12}$$

where δ_a and δ_b represent local mismatch between the paths in multiplexer, A and B, respectively. It is clear that this is only a local mismatch rather than mismatch

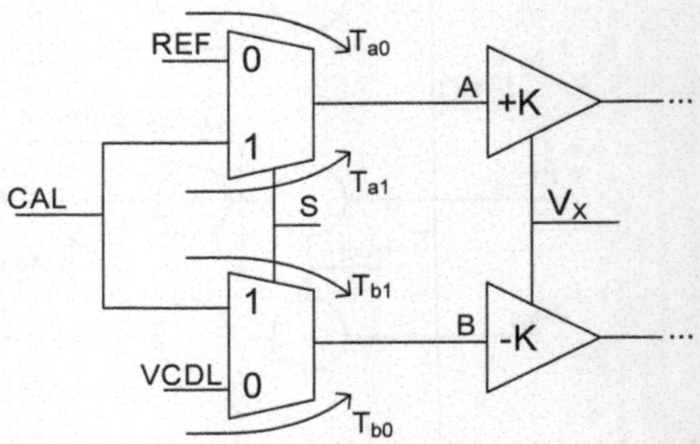

Fig. 4.17 Multiplexer path mismatch

between different multiplexers. These paths can be spaced closely together such that spacing dependent effects (e.g., doping gradients) have little influence. The remaining offset results from random variations between the two paths in the multiplexer. Thus, this variance could be made very small by using large transistors to reduce the local mismatch. However, this would again increase power consumption of the entire timing macro or require buffers that, at their turn, generate offsets. In this design minimum sized cells are used resulting in the highest offset possible to prevent increasing power consumption in the TDC and DLL.

The remaining timing variation between these paths is lower than the internal offset of the original phase detector such that invoking a calibration is useful. The core phase detector in this design had a 1σ offset variation of 1.7 ps with a zero mean offset. Therefore, only process variations have to be canceled with this technique. With the proposed calibration technique, the variation on this offset can be reduced to 500 fs. This remaining offset results from timing mismatch described earlier showing that a significant improvement is achieved using this calibration method.

If there is an imbalance between the calibration loop's charge-pump currents, the limit cycle will not be symmetrical anymore. This results in an offset shift from the inner loop. A decrease in the charge-pump current decreases the limit cycle timing error amplitude and thus reduces the static offset from this mismatch.

Concerning the delays of the multiplexers, in newer CMOS technologies, random variations increase. However, it is the absolute value of the timing mismatch between the multiplexer paths that is of concern in this circuit. Since the gate delay becomes smaller in newer technologies, the offset will reduce as well. Therefore, scaling will be advantageous for this technique.

Since a symmetrical layout is preferred at the input paths of the phase detector, identical cells are used for both delay cells to improve the matching. Figure 4.18 shows the implementation of the delay elements. A two-stage delay allows a configurable voltage to delay gain depending on the connection of the cell. The first stage is a nmos current starved delay cell. High nmos control voltages result in a faster stage and thus negative voltage to delay gain. The second stage is complementary in which a pmos current starved delay controls this stage resulting in a positive voltage to delay gain. An overall positive gain is obtained by connecting the control voltage to the pmos stage and enabling the nmos stage at full speed. The

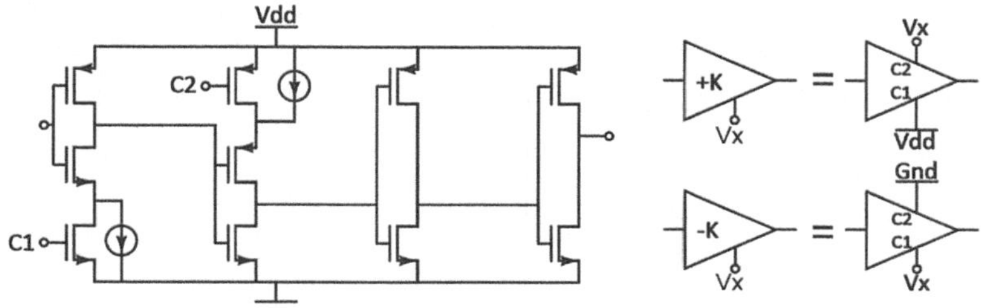

Fig. 4.18 Input delay cell implementation

negative gain is complementary enabling the pmos stage at full speed. Since the phase detector is positive edge sensitive, only positive edges have to be delayed. This reduces the circuits overhead to delay both edges equally.

The circuit allows a tuning of ±20 ps which is enough to compensate for the internal offset of the phase detector. Since this circuit acts as a differential tuning topology, the circuit will stabilize at the point where roughly both delays are equal because the offset to compensate for is much smaller than the tuning range of the delays. The control range of the cell varies from 0.25 V up to 0.65 V. This ensures both current sources of the charge-pump operate in saturation in a 0.9 V 40 nm CMOS technology.

4.4 Experimental Results

The reference frequency of the DLL can be set from 800 MHz to 2.4 GHz. Depending on the applied frequency, the LSB bin width can be programmed from 14 ps down to 4.8 ps using the five times resistive interpolation DLL. The power consumption ranges from 2.4 to 4.2 mW, respectively. The measured control voltages of the VCDL for different frequencies are shown in Fig. 4.19.

Different samples were measured to verify the linearity of the TDC due to local process mismatch. Global mismatch variations that change the entire chip's performance are compensated by the DLL's control loop which only requires enough tuning range. However, local variation between different delay cells of the VCDL and capture time difference of the flip-flops result in the remaining non-linearity. The linearity measurements are shown in Fig. 4.20 for a nominal 7.8 ps bin-width. The phase difference between the DLL's reference clock and hit signal was varied over 2π to characterize the entire range of the TDC using a

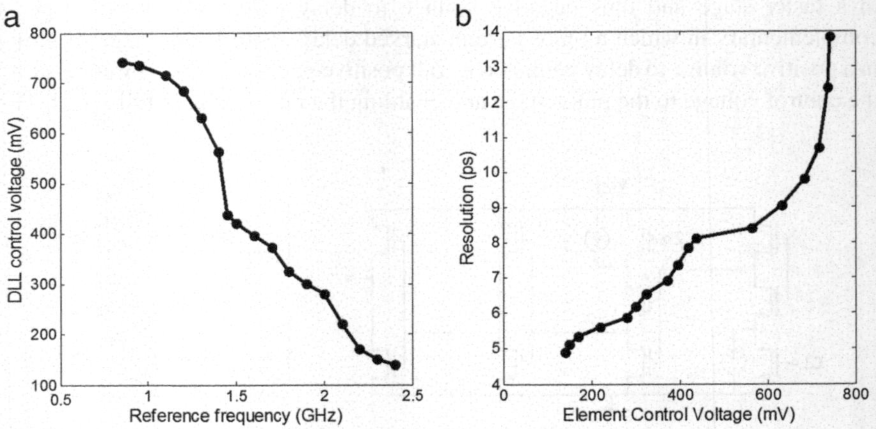

Fig. 4.19 (**a**) Measurement of the DLL lock voltage for different frequencies. (**b**) Calculated gate delay for different control voltages

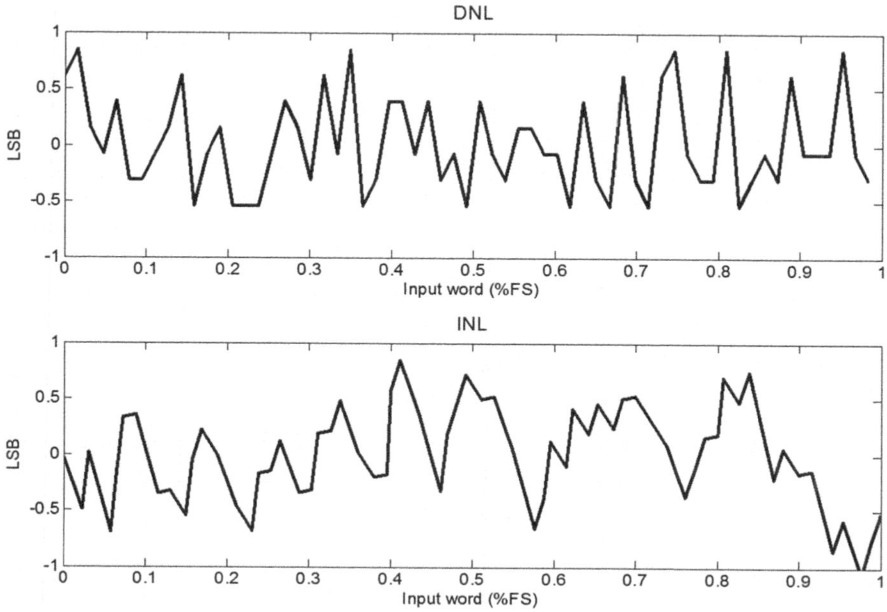

Fig. 4.20 DNL and INL measurement

code density test. The appearing pattern in the DNL results from non-uniform RC loading in the resistive interpolation due to additive layout parasitics. However, this could be removed with calibration. While this range covers only 6 bits, it can be extended for larger dynamic ranges with a reference counter where only long-term accumulated jitter of the reference clock limits the range. This figure shows the INL and DNL performance of the TDC expressed in LSB between two interpolated nodes. Different measured chips show similar results. A 0.43 LSB DNL rms and 0.4 LSB INL rms is measured.

Figure 4.21 shows the startup speed of the DLL as a function of the applied bias current of the charge-pump of the PFD. A higher current does not result in an excess control voltage ripple as would be the case with a single BBPD detector. For larger charge-pump currents, the speed does not increase anymore because of saturating current sources. Beyond the time reported in Fig. 4.21, the PFD enters its deadzone window and shuts down automatically until the DLL is disturbed. When shutting down, no current flows and no offset is introduced by the PFD's charge-pump. A standard charge-pump architecture is used without the need for any intense analog design methods to cope with clock feed through, charge injection, or output impedance mismatch.

The offset measurements were performed by measuring both on-chip signals REF and VCDL at the phase detector input within a closed loop DLL. Since very small sub-picosecond offsets have to be measured, only a single output buffer is

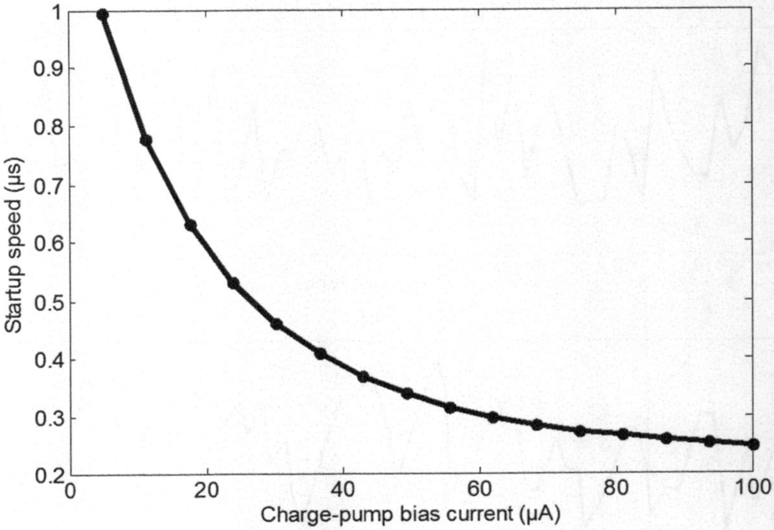

Fig. 4.21 DLL startup speed

used to prevent delay mismatch between two chained output buffers. An extra on-chip multiplexer selects the REF or VCDL signal to the single output buffer that brings the clock signals off chip as is shown in Fig. 4.22a. A tapered 50 Ω output driver is used to minimize capacitive loading to the minimum sized multiplexer. The selection of the multiplexer is performed using an off-chip select signal. The offset between REF and VCDL is thus measured by measuring the delay difference between those signals relative to the input reference clock from the signal generator. In this way, signals REF and VCDL share the same output driver and PCB lines and thus no extra timing mismatch occurs in this path. Since there is an extra multiplexer in the output path, a higher offset variation is measured than exists on the chip because of the mismatch in the output selection multiplexer. The offset of multiple chips was measured resulting in a 1σ offset variation of only 0.5 ps where all circuits are implemented using minimum sized logic gates to prevent signal buffering. The mean of the offset measurement (8 fs) was much lower than the 1σ deviation to be relevant. This value corresponds to a nominal supply voltage of 0.9 V and includes the delay mismatch at the multiplexer output paths. The measured value precisely matches with SPECTRE Monte-Carlo spice simulations. Simulations have shown that this is more than three times better than using the same phase detector without mismatch calibration. In this setup, a sample size of 14 has been taken. Figure 4.22b shows the 1σ offset variation at different operation conditions of the phase detector. As shown in Fig. 4.22b, the offset variation over different chips varies significantly with the supply voltage. This result can be explained from the multiplexer delay. Since the delay variance of the multiplexer defines the offset of the phase detector, faster gates at the input result in smaller offsets because the

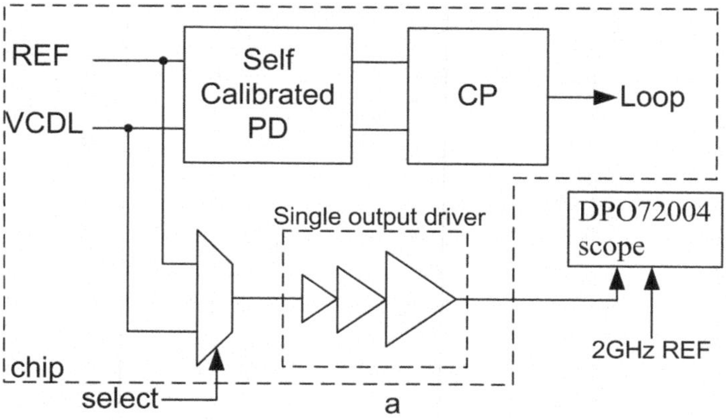

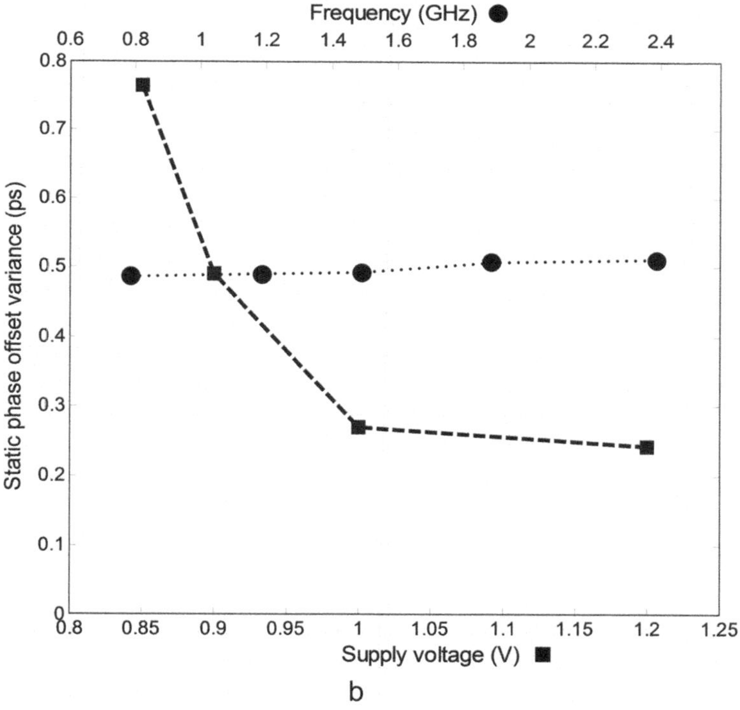

Fig. 4.22 (**a**) Offset measurement setup. (**b**) Offset variance as a function of supply voltage and Offset variance as a function of reference clock frequency for the proposed phase detector

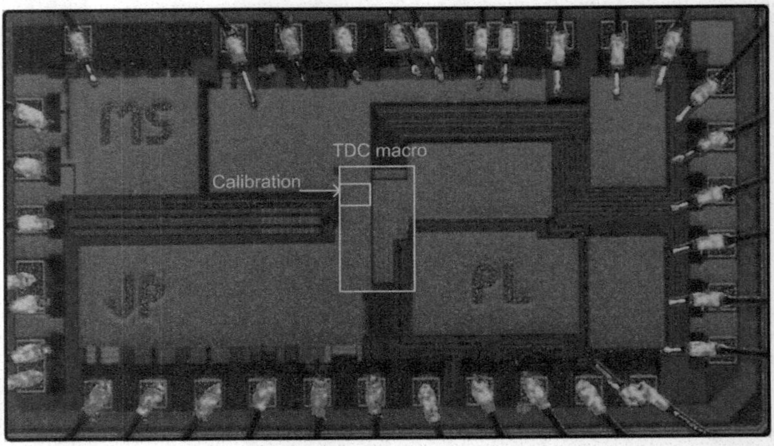

Fig. 4.23 Die photograph

offset of the multiplexer is proportional to its delay and the signals' slew rates which improve at higher supply voltages. Figure 4.22b also shows the offset variation as a function of different clock frequencies. It is clear that there is almost no frequency dependence on the offset of the phase detector because the gate delay or slew rates remain equal. Figure 4.23 shows the die photograph of the TDC that has been manufactured in a 40 nm CMOS technology which is mounted on a test PCB as is shown in Fig. 4.24. The TDC macro includes the DLL with self-calibrated phase detector. Since approximately 50% of the area in this single-channel test structure is occupied by the DLL, it is clear that only a small fraction of the area is occupied by the calibration circuitry.

A table to compare the performance to relevant work is shown in Table 4.1. A performance summary is shown in Table 4.2.

4.5 Conclusion

This chapter presented an experimental TDC which targets nuclear applications. The TDC is based on a DLL to stabilize the delays in the delay line against PVT variations and TID effects. A hybrid phase detector is presented which combines a PFD and BBPD. This method provides a high startup speed, low static offsets, and low jitter. The improved startup speed enhances the recovery time after an SEU on the circuit. A low offset, self-calibration circuit is presented which compensates for static offsets in the BBPD. The circuit operates in 2 states to self-measure its offset and applying this to the measurements of the detector.

A resolution of 4.8 ps has been achieved with a power of 4.2 mW and a 0.43 LSB rms and 0.4 LSB rms DNL and INL, respectively. The self-calibrating phase

Fig. 4.24 Picture of the PCB with power and IO interfacing

Table 4.1 Comparison of the performance of this TDC

	[129]	[68]	[130]	[131]	This work [132]
Technique	Time amp.	Passive interp.	Delay line	Pipeline	Passive interp.
Sample rate (Msa/s)	10	180	26	300	150
Resolution (ps)	1.25	47	17	1.76	4.8
INL (LSB)	3	1.2	0.9	1.9	0.84
Dynamic range (bits)	9	7	6	10	6
Power (mW)	3	3.6	6.9	115	4.2
Core area (mm^2)	0.6	0.02	0.01	0.88	0.04
Technology (nm)	90	90	90	130	40

Table 4.2 A summary of the performance of the TDC in this work

Process	40 nm CMOS
Supply voltage	0.9–1.2 V
Resolution (bin width)	14–4.8 ps
Operating frequency	0.8–2.4 GHz
Static phase-offset	0.5 ps
Startup speed	250 ns
Power consumption	
@0.8 GHz (14 ps)	2.4 mW
@1.5 GHz (7.8 ps)	3.0 mW
@2.4 GHz (4.7 ps)	4.2 mW
Dynamic range	6 bit
INL (LSB)	0.84 max, 0.43 rms
DNL (LSB)	0.96 max, 0.40 rms
Core area	156 μm × 256 μm

detection circuit reduces the static offsets down to 500 fs which is three times better than an uncalibrated circuit.

The dynamic range is limited by the reference clock period but it can be extended through a counter, running at the reference frequency. This topology is now being processed in a revision of this circuit to improve the linearity and process variations on the design.

Chapter 5
Low Jitter Clock Generators

5.1 Introduction

This chapter will discuss design and implementation issues for low noise clock generators in nuclear applications. This clock generator is associated with high resolution Time-to-Digital Converters. TDCs are required in several of today's high-energy physics experiments such as CMS and ATLAS at CERN. In such applications, it is important to measure the time difference between different hit channels (originating from the detector circuits) of the experiment with single-shot precision rather than oversampling converters. The TDCs, required in those applications, are typically multichannel circuits with channel counts ranging up to 64. The TDCs are based on a timing generator which is a DLL locked to a high frequency reference clock. In this SoC, the frequency of the DLL is targeted to run at 2.56 GHz. Since the timing resolution of the TDCs is aimed in the picosecond range (3 ps bin size), the reference clock stability of the DLL becomes critical in terms of jitter or phase noise. Ideally the DLL would receive a very clean high speed clock with no uncertainty on the clock edges but in reality there will be jitter. Since this TDC will be used in LHC experiments, the circuit will use the 40 MHz reference of the accelerator which is multiplied by 64 in the PLL.

Figure 5.1 shows the complete architecture of the TDC in which this PLL is integrated.

The SoC consists mainly of 5 different blocks. The first 3 full custom blocks are: (1) A PLL clock generator which generates a low noise, high speed 2.56 GHz clock. This circuit is the subject of this chapter (2) A DLL which generates the timing bins of the TDC, and (3) 64 hit channels which sample the DLL timing generator.In the

J. Prinzie et al., *Radiation Hardened CMOS Integrated Circuits for Time-Based Signal Processing*, Analog Circuits and Signal Processing,
https://doi.org/10.1007/978-3-319-78616-2_5

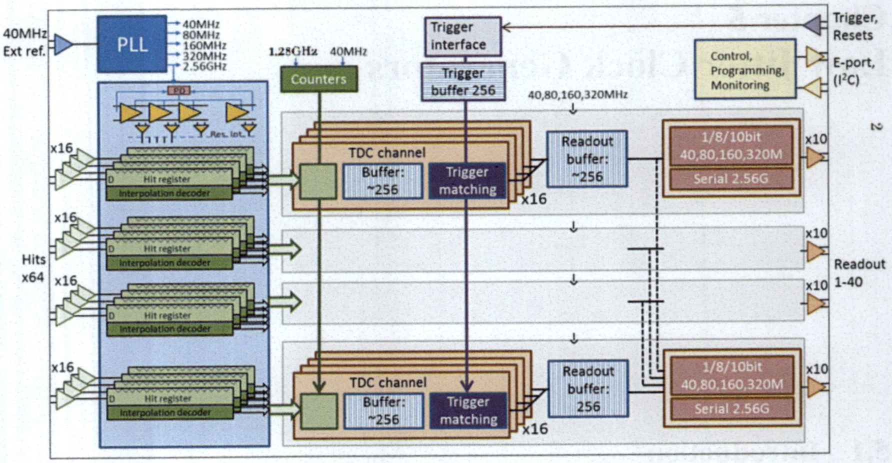

Fig. 5.1 Architecture of the 64 channel picosecond accurate TDC SoC

Table 5.1 PLL clock synthesizer requirements

Output frequency	2.56–3.2 GHz
Multiplication factor	64
Integrated rms jitter	<1 ps
Power consumption	<20 mW
Reference frequency	40–50 MHz

back-end, (4) a fully synthesized digital processing and buffering system decodes and stores the data until the chip is triggered. The data is interfaced through (5) the readout IO with 40 Mbps–2.56 Gbps interface speeds.

To meet the specifications of the SoC, first, the circuits have to be robust and compatible with industry standard voltage and temperature variations [133]. Since the systems will be produced in medium-size quantities, yield due to process variations should be sufficient meaning the design has to meet the required performance in all PVT corners.

The performance requirements of the PLL are listed in Table 5.1.

In this chapter, in the first part, the architecture of the PLL is presented which has been used to implement a prototype chip. Secondly, radiation effects on PLLs and its components are discussed. Then, the oscillator and phase detector are discussed. Finally, the measurements of this circuit are discussed. The next chapter is dedicated to radiation experiments on this PLL. The main motivation for this prototype was to understand the robustness and radiation issues encountered in charge-pump PLLs, furthermore, the interest rose to compare LC-oscillators with ring oscillators for its radiation hardness, both TID and SEE.

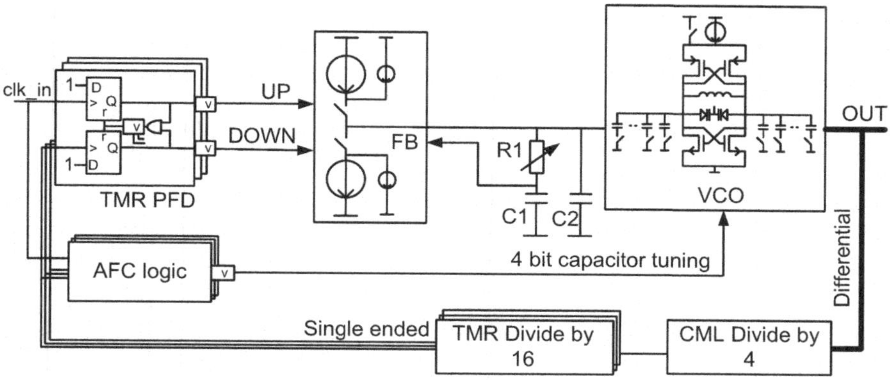

Fig. 5.2 Architecture of the prototyped PLL

5.2 PLL Architecture

The charge-pump PLL which has been prototyped in this work is shown in Fig. 5.2. The PLL has an LC-tank oscillator with a wide tuning range which is discussed in more detail in Sect. 5.4. A divide-by-64 is implemented in two stages: a divide-by-4 which is designed in CML, followed by a divide-by-16 in which CMOS standard cells are used. The latter has TMR protection against single-event radiation effects. The CML prescaler was used to ensure sufficient timing overhead in the standard cell logic as the speed may degrade due to TID effects. A new TMR PFD architecture is proposed in this research which provides asynchronous logic protection against SEU soft errors. The phase detector is discussed in Sect. 5.6. Finally, since the PLL has to work with a relatively large input frequency range, the oscillator has extended discrete tuning bands which can be digitally controlled [134, 135]. An all-digital search algorithm is used after Power-On Reset (PoR) to find the optimal band in the VCO. This algorithm is discussed in Sect. 5.7 [136].

The circuit has been processed in a commercial 65 nm CMOS technology with a 1.2 V power supply.

5.3 Common Radiation Effects on CMOS PLLs

5.3.1 Charge Pump/Loop Filter

5.3.1.1 TID

Recall from Chap. 3 that the loop dynamics (bandwidth and damping) depend on the value of the charge-pump current. Furthermore, the reference spurs and also the static-phase offset depend on the imbalance between the up and down currents.

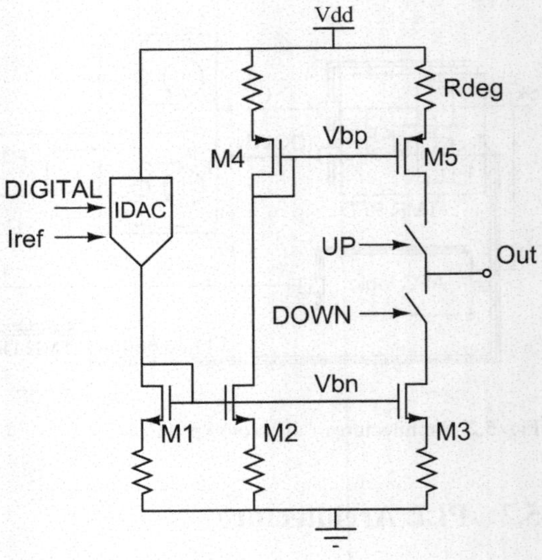

Fig. 5.3 Full charge-pump with degeneration resistors

Since TID effects degrade the current conduction and threshold voltage of the charge pumps, they may introduce degraded dynamics in the PLL.

However, charge-pumps are in general relatively robust in first order against small variations in the currents of the devices, it is more a matter of matching. Figure 5.3 shows a schematic of the charge-pump from this design. If the DAC generates a constant current, independently of the dose, the voltage V_{bn} is

$$V_{bn} = R_{deg} I_{cp} + \left(V_T + \sqrt{\frac{2I_{cp}}{\mu C_{ox} W/L}} \right) \tag{5.1}$$

Due to total dose effects, the current drive capability of M1–M5 decreases. But, since it is configured as a current mirror, the overdrive voltage will increase. Therefore, first order effects do not degrade the charge-pump current. However, 2 issues occur at the output. Firstly, since V_{bn} and (V_{bp}) increase (decrease), the relative V_{ds} matching between M4/M5 and M2/M3 reduces which results in an increasing current imbalance. Secondly, since the overdrive voltage increases, the allowed swing at the charge-pump's output reduces as the transistors enter the triode region sooner. For example, if the current in the devices degrades by 10%, the overdrive voltage increased by only 5.4%. This simulation is shown in Fig. 5.4

These effects cannot be fully mitigated but with proper design, they can be minimized to be secondary degradations. Transistors M1–M5 are DC current sources and can have a large device area (W and L). This is advantageous for the output resistance and also for TID effects. Recall from Chap. 1 that devices with a size larger than 500 nm degrade only by 10% or less which makes the above

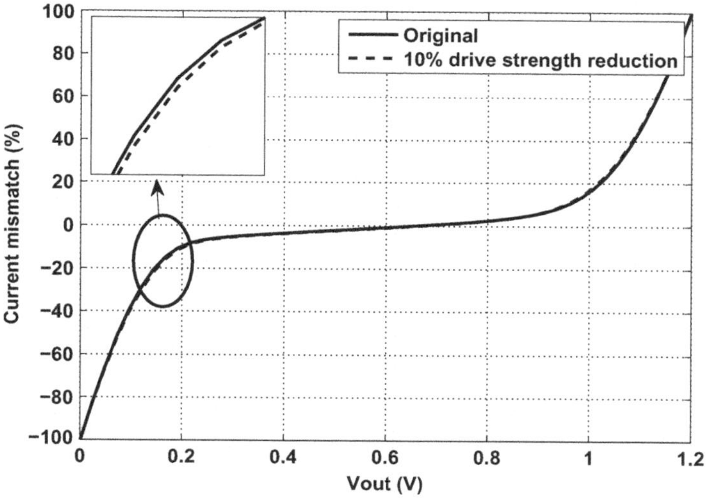

Fig. 5.4 Simulation of the current mismatch before and after 10% degradation

effects reasonable in CMOS PLLs. Generally ELTs are not recommended for these transistors since it is not practical to design an ELT transistor with a length larger than 100 nm.

The switches which are controlled by up and down require some special attention. The voltage drop across these devices should be as small as possible since this reduces the voltage headroom in the output branch. However, too large devices increase charge injection due to the channel and capacitive feed through, thus relatively small transistors with minimum length are preferred. They however degrade significantly with total dose effects such that ELT devices are mandatory at the output. A detailed study on the TID effects in charge-pumps can be found in [137].

5.3.1.2 SEE

Single-Event Effects primarily affect the charge-pump in two ways. The first circuit which is affected is the biasing circuit which generates the bias voltages V_{bp} and V_{bn}. An SEU on these nodes can disturb the bias voltage which in turn alters the control on the PLL. If the SEE affects the node V_{bn}, also V_{bp} will be adjusted such that the loop gain momentarily decreases. If, however V_{bp} is affected only, the up current will momentarily change. In the first case, the loop is less stable for some period of time. In the second case, there is an additional current mismatch. The effect of these errors will largely depend on the bandwidth at the nodes V_{bp} and V_{bn} and on the bandwidth of the PLL. However, this issue can be easily resolved when these nodes are decoupled.

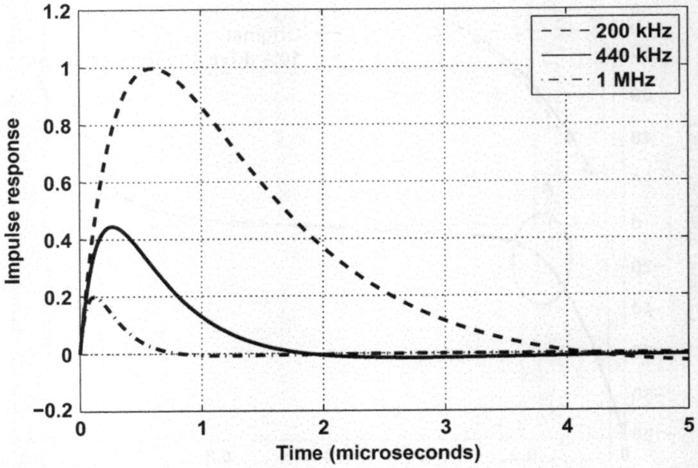

Fig. 5.5 Impulse response to an SEE on the loop filter with a constant capacitance for a 200 kHz, 440 kHz, and 1 MHz bandwidth

The second issue is the output node which is connected to the loop filter. The SEE can be modeled as a dirac impulse parallel to the charge pump. The frequency domain transfer function that this impulse produces to the output of the PLL is

$$NTF(s) = \left| \frac{2\pi}{I_{cp}} N \frac{\omega_n^2 (1 + \frac{2\zeta}{\omega_n} s)}{s^2 + 2\zeta \omega_n + \omega_n^2} \right|^2 \tag{5.2}$$

Intuitively, a larger loop capacitor will result in less voltage shifts due to an injected current impulse. Thus, for a given loop bandwidth, larger capacitors will create less phase jumps in the PLL [138]. However, the loop bandwidth itself also influences the phase deviations due to an impulse. Figure 5.5 shows the impulse response for a PLL system with constant loop capacitance with a bandwidth of 200 kHz, 440 kHz, and 1 MHz. It is clear that lower bandwidths suffer more from an SEE compared to large bandwidths. This can be easily explained as follows. Given a loop capacitance, the voltage shift, generated by an impulse will be constant which results in a frequency step in the VCO. As a result, the phase will drift away. The PLL will compensate for this drift but will do this more slowly at low bandwidths which allows the phase to accumulate more than at high bandwidths.

In [139], a layout technique is proposed which uses charge sharing between the up and down switched to partially cancel the transient currents. Another technique to reduce the effects of single-event in charge-pumps was to completely remove the current sources and use resistors instead [140]. This allowed the PLL to more quickly recover from a single-event transient. However, this comes at the cost of a larger reference spur due to the small output impedance of the charge-pump.

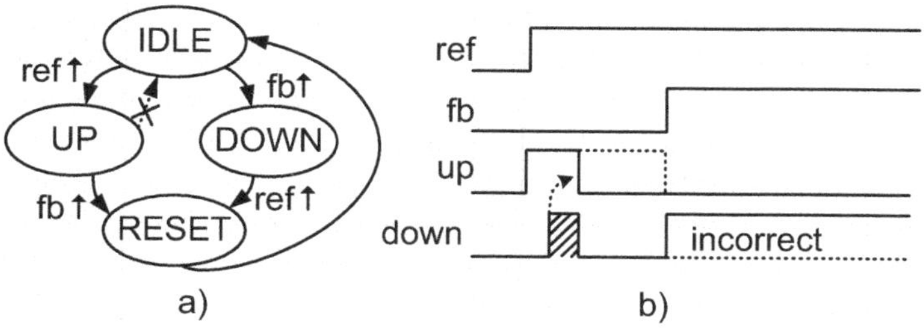

Fig. 5.6 (**a**) FSM from a PFD. The dashed line indicates an incorrect state switch. (**b**) Signals from the PFD when an SEU occurred at the down signal

5.3.2 Phase Detector

A PFD is a simple 4-state FSM (Finite State Machine). When Idle, it expects an edge on the reference or feedback clock to move to its next state which should be shortly followed by an edge on the other clock as shown in Fig. 5.6a. Due to an SEU in the registers, the state of the FSM can be incorrectly switched which results in a cycle slip in the PLL. This means that the PLL loses lock and needs to acquire lock again which takes a significant amount of time. An example is shown in Fig. 5.6b. An error occurs on the down signal which results in an incorrect reset of the PFD. When later, the feedback clock arrives, this should reset the PFD but does not, which results in a 360° phase shift, seen by the PFD. TMR should be employed to prevent these errors. However, since the circuit has a memory, the TMR should be self-correcting and not a parallel structure with 3 identical phase detectors with a voter [141].

Since the PFD typically works at low frequencies (40 MHz) in this case, there is no issue resulting from increasing delays in the logic at high radiation doses.

5.3.3 Divider

A divide-by-N counter will contain $\log_2(N)$ bits in its digital logic. Each of these bits can be upset due to an SEU in the logic. The phase deviation will depend on the bit which is upset. If the MSB is upset, it can create 180° phase errors, while if the LSB is upset, the phase error is only $180°/\log_2(N)$. This is a severe error, even if an LSB is considered. Therefore, the logic must be protected against SEUs. This can be easily done with TMR.

The TMR however has a few drawbacks. Firstly, the logic typically runs at a relatively high frequency where triplication with additional voters requires 3–4 times more power compared to a regular design and consumes more than 3 times the area. Secondly, a speed penalty occurs. Due to the extra voting logic, the combinational logic introduces more delay which reduces the maximum frequency.

Since the design may run at high frequencies, TID effects can become an issue. Especially if foundry CMOS standard cell libraries are used to synthesize the dividers since they use relatively small devices. Either, a large timing overhead is required or ELT standard cells should be used. Generally, ELT cells cannot be as small as minimum sized standard cells due to geometric process rules. Therefore, its power consumption is typically much larger compared to traditional standard cells. ELTs should therefore only be used when the timing slack of the digital design is small. In case of a large slack, non-ELT cells can be used.

5.3.4 VCO

5.3.4.1 TID

LC-tank oscillators are intrinsically well protected against total dose effects in the silicon. Recall that the oscillation frequency of the circuit is set by

$$\omega_0 = \frac{1}{\sqrt{LC}} \tag{5.3}$$

in which L depends on the geometry of the inductor and C mainly on the capacitances in the oxide. These parameters do not change with radiation which means that the frequency will not drift due to total dose effects. One possible threat lies in the loop gain of the oscillator. To ensure the oscillator starts, $G_m R_p > 1$ which means that G_m should be sufficient. Due to accumulated charges in the silicon, the total transconductance of the devices decreases. Therefore an additional safety margin is required to ensure the oscillation will remain after irradiation. Since the devices in an LC tank are typically large, an expected reduction in the G_m of less than 30% is expected without any special layout techniques.

The situation for ring oscillators is different. In these circuits, the oscillation frequency is set by

$$f \sim \frac{I}{2 \cdot N \cdot C} \tag{5.4}$$

in which I is determined by the transistors and will significantly degrade with accumulated dose. ELT devices are then mandatory in the circuit.

5.3.4.2 SEE

The sensitivity of a VCO to SEEs can be divided into three parts. (1) The biasing circuitry, (2) the tuning circuit, and (3) the core of the oscillator as shown in Fig. 5.7.

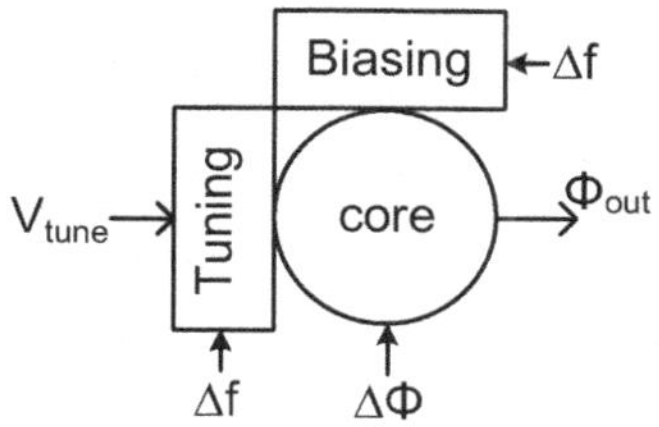

Fig. 5.7 Three different sensitive parts of a VCO

The biasing circuitry generates all bias voltage and currents in the circuit. For example, in a ring oscillator, the "zero-input" current is an example of such circuit which sets a minimal current through the voltage controlled delay cells. If this current is affected by an SEE, the frequency changes. These effects can be easily mitigated through decoupling, similar to the effects in the charge-pump biasing current mirrors.

The tuning circuit of the oscillator ensures that the frequency is adjusted based on the input voltage of the VCO. For ring oscillators, this is a simple current mirror with an input transconductance device which converts voltage to current. Similar to the biasing structure, an SEE on these nodes will cause a frequency shift in the oscillator. In an LC tank, this tuning is done through a varactor which can capture charges in the active areas and cause frequency jumps. Mitigation through capacitive loading is difficult in these parts of the circuit since a moderate speed is required. The bandwidth of these circuits should be sufficiently large such that the poles in the PLL loop do not degrade the phase margin.

Finally, the core of the oscillator is sensitive to collected charges as well. While the previous effects result in frequency jumps, charges in the oscillator contribute to phase steps. The same mechanism can be applied as was done to calculate the phase noise using the Impulse Sensitive Function in Chap. 3. Recall that the phase impulse response to a dirac current impulse is

$$h_{\phi}(t,\tau) = \frac{\Gamma(\omega_0 \tau)}{q_{max}} u(t-\tau) \tag{5.5}$$

From this equation, it is clear that a larger q_{max} in the oscillator contributes to smaller phase jumps [142]. If $\Gamma(\omega_0 \tau)$ can be reduced, it will also result in a reduction of the phase amplitudes. If the charge is injected when the signal is at its maximum, the net phase jump will be minimal (the ISF will be small in this region). In contrast, when this happens at a zero-crossing, the effect will be maximal. In ring-oscillators though, multiple (N) nodes are present for which the phase of the individual ISF is shifted by $2\pi/N$. Finally, the phase deviation can be reduced by increasing q_{max}. This involves a larger capacitance and more power in the oscillator.

If ring oscillators are compared to LC-tanks and analyzed through q_{max}, it becomes clear why LC-oscillators are much more tolerant against SEEs. In an LC tank, the parallel resistance of the tank is

$$R_p = \frac{Q}{\omega_0 C_{tank}} \tag{5.6}$$

Thus, the required power consumption for a charge $q_{max} = CV$ peak-to-peak (differential) voltage V is

$$P_{LC} = IV_{dd} = \frac{V \cdot V_{dd}}{\frac{4}{\pi} R_p} = \frac{V\omega_0 C}{\frac{4}{\pi} Q} V_{dd} \tag{5.7}$$

In ring oscillators, the power in the cells for a charge $q_{max} = CV$ is

$$P_{ring} = IV_{dd} = NCV_{dd}Vf_0 \tag{5.8}$$

The ratio between both power consumptions, for the same q_{max} is

$$\frac{P_{ring}}{P_{LC}} = \frac{2NQ}{\pi^2} \tag{5.9}$$

from which it can be concluded that the charge in the LC-tank is larger compared to the charge in the node of a ring oscillator, for the same power consumption. Especially when N grows. This can also be explained through the quality factor which is defined as the energy stored during a cycle and the energy dissipated which is larger for LC-tanks.

Different architectures were used before to attempt to harden the oscillator against single-event effects. In [143] frequency to voltage feedback was used on the VCO control voltage to compensate for any disturbances on this signal. In [144], the bias current source in an LC-tank oscillator was improved with a series resistor which reduced the effects of a single-event on the bias current. In this design, however, a large capacitor was used to prevent single-events on the bias currents.

Note that technology scaling does not improve the single-event sensitivity of ring oscillators (and digital circuits in general) since the phase displacements are determined by the ISF and the total capacitance and voltage which reduces with shrinking devices [145].

A detailed experimental discussion is given in the next chapter in which a ring oscillator and LC-tank oscillator are compared in terms of SEE and TID effects. In Sects. 5.4 and 5.5, the circuits which are used in these experiments are described.

5.4 Integrated LC-Tank Oscillator

The VCO has been implemented as a 5-transistor -Gm cell with pmos current source to reduce 1/f phase noise as is shown in Fig. 5.8. Since this design has to meet

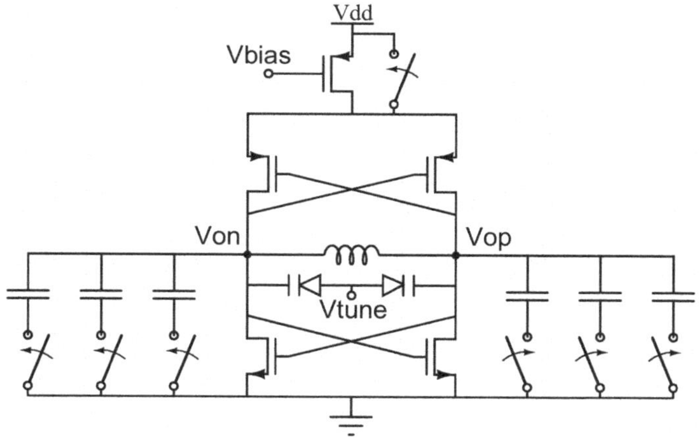

Fig. 5.8 Schematic of the implemented LC-tank oscillator

different application requirements (in terms of phase noise and supply rejection), the current source of the VCO can be bypassed to the supply. When the current source is enabled, the single-ended swing of the oscillator is limited to 0.6 V (Vdd/2) to keep the drain of the current source in saturation for good power supply noise rejection (PSRR). In applications where the PSRR is less important, the current source can be bypassed to the supply to improve the signal swing and decrease the phase noise of the oscillator. The bias voltage of the pmos current source is generated from a supply-independent bias generator.

The varactors are implemented with MOSCAPS (Metal-Oxide-Semiconductor Capacitors) and a tuning overlap of more than 50% is ensured between different bands in the VCO. These varactors are directly coupled to the oscillator. When the supply switch is on, the DC voltage at the output nodes is approximately $V_{dd}/2$ such that the tuning range of the varactors is maximized. If the VCO operates with the current source, the DC output voltage will be 0.3 V which results in a reduction of the tuning range (or overlap between bands). An optimal inductor of 700 pH has been chosen in relation with varactor and capacitor-switch losses using the design methodology as discussed in Chap. 3. The inductor was chosen from the PDK such that its equivalent parallel resistance matches required swing with a geometry that provides the highest quality factor (smallest Rs). This inductance minimizes the phase noise for a 5 mW power budget at this frequency range.

The switches of the capacitor banks are implemented with two ground switches and one (dominant) lateral switch as is shown in Fig. 5.9a. The lateral switch improves the quality factor of the capacitor for the same parasitic capacitances to ground compared to ground switches due to a virtual ground in the symmetric signals. Due to this virtual ground, the series resistance of M2 is divided across both capacitors as is shown in Fig. 5.9b. The small switches to ground are used for

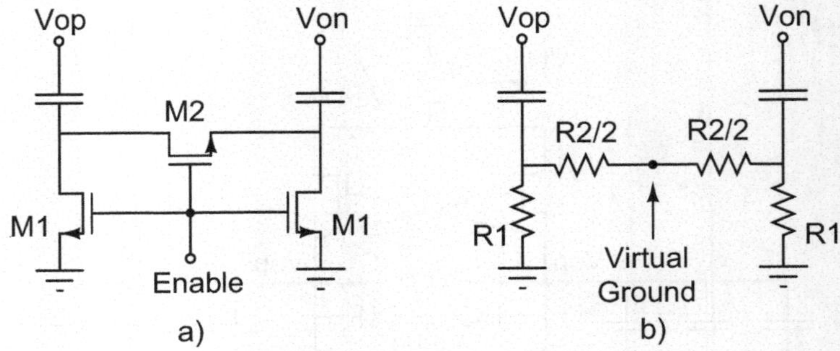

Fig. 5.9 (**a**) Implementation of the digitally switched capacitor banks. (**b**) Switch resistance when activated

Table 5.2 Parasitic inductance in the connections to the digitally controlled capacitors

d(μm)	L(pH/μm)	C(fF/(μm))
20	0.3	0.210
60	0.39	0.202
600	0.56	0.197

biasing to maximize the overdrive voltage of the lateral switch. The capacitor banks in this design are implemented with MOMCAP capacitors. These capacitors have an improved quality factor compared to MOSCAPS but have slightly higher process variability. Therefore, the automatic frequency selection algorithm is added to select the correct setting.

A careful analysis is required when the connection to the switched capacitors is designed. A chip photograph is shown in Fig. 5.10 from which it is clear that the traces which connect the inductor to the capacitors are relatively long. Since the inductance is small, they become significant and may change the oscillation frequency. The inductance which is series connected with the switched capacitors lowers the equivalent impedance of the bank of capacitors, which are added to the tank. Therefore, it appears to the inductor that the capacitor is larger than implemented. In this design, the symmetry of the currents can be exploited by spacing these connections closer such that the total differential inductance is reduced. For this, 3D EM (Electromagnetic) simulations were done on the layout to model this connection in the post-layout simulations. Table 5.2 shows a simulation for the geometry in Fig. 5.10 with a spacing between the long tracks of 20 μm up to 600 μm with its according equivalent inductance and capacitance. The simulations show that the inductance per unit length decreases when the spacing decreases while the capacitance remains relatively constant. This capacitance is dominated by the substrate coupling of this track and not by the edge coupling of the two tracks.

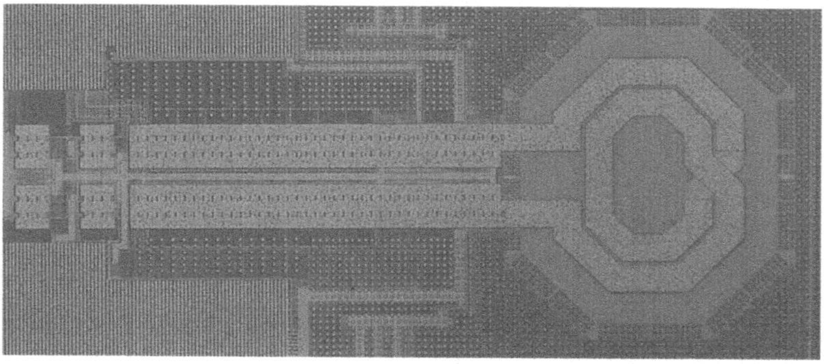

Fig. 5.10 Die photograph of the LC-oscillator

5.5 Ring Oscillator as a Comparison

To compare the radiation sensitivity against LC-tank oscillators, a ring oscillator has been added to this prototype. Each oscillator on the device has a full PLL controlling its phase such that botch circuits can be tested simultaneously. However, the gain of the ring oscillator (MHz/V) is much higher compared to the LC-tank oscillator. Though the digital logic is the same, the charge-pump current is different to cope with the higher gain in the ring oscillator.

The oscillator was implemented and designed with the same power consumption as the LC-tank oscillator to allow a fair comparison. Four delay cells were used to achieve an oscillation frequency centered at 2.56 GHz. The schematic of a delay cell is shown in Fig. 5.11a. M3a operates as a current source and M2a–M2b is a pmos input pair. The input pair is loaded with a "Maneatis" load which consists of two identical nmos devices M1a/M1b from which one is connected as a diode and the other as a linear load. The idea behind this structure is that the resistance of this load is more symmetrical around its switching point [146].

To control this delay cell, a biasing circuit is used which generates the required voltages, depending on the control voltage input of the VCO as is shown in Fig. 5.11b. In the first branch, the voltage is converted to a current in M4, from which $biasp$ is generated. This current is mirrored to the second branch through M3b/M3c where $biasn$ is generated. M2c acts as a dummy for matching to the delay cells and a replica of M1a/M1b is used to generate the voltage in M1e/M1f. In real Maneatis cells, the linear load is set by an opamp which fixes the minimum output voltage. However, in this design, no amplifier is used to reduce the SEE cross section for the circuit. The circuit has a relatively stable output voltage swing among different control voltages due to the replica bias that generates $biasn$. The additional resistor is used to ensure a minimal current flowing in the delay cells when the input voltage is zero.

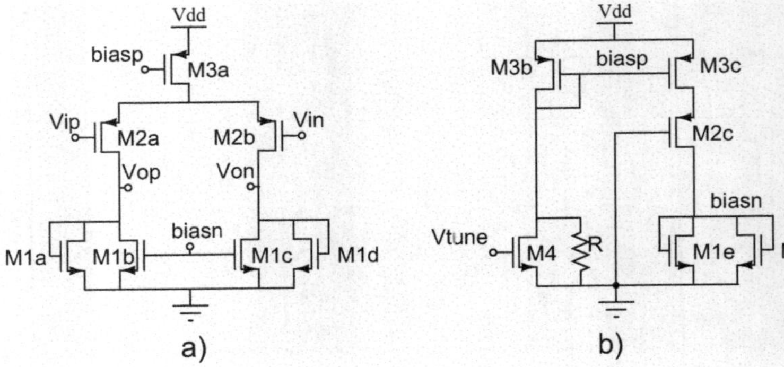

Fig. 5.11 (**a**) Delay cell schematics. (**b**) Input tuning circuit which generates the required bias voltages

The SEE sensitivity of the oscillator increases if more stages are used for a constant power consumption. For an even amount of stages, 4 is the minimum that can be used for an oscillation which is used in this design.

5.6 TMR Phase-Frequency Detector

A Triple Modular Redundant phase detector is implemented as shown in Fig. 5.12 to reduce the sensitivity against Single Event Upsets. It consists of three identical cells, each representing a single PFD. The flip-flops automatically load a logic 1 and are implemented with two NOR based CMOS latches. Each PFD calculates its own UP, DOWN, and reset signals but does not directly use the reset, generated by the AND gate, to reset the PFD. Instead, a majority voted representation of the reset is calculated from the three reset signals of the PFDs. In this way, the PFD automatically recovers to the correct state if a Single-Event Upset has occurred on one of the flip-flops. If no cross-coupled reset is used, the state would never be corrected. Each PFD has its own voter for the reset signal to prevent an incorrect reset in case of an SEU on the voter. In this way, an SEU on the reset voter can only incorrectly reset a single PFD which will be corrected in the next cycle. The signals UPA—UPC and DOWNA—DOWNC are voted at the output of the phase detector and are routed to the charge pump. TMR can be further ensured to the charge pump by placing 3 series switches to control the current of the charge pump. However, this is not implemented in this design. A single error on the phase detector does not result in an incorrect behavior that would result in PLL cycle slips. Double errors (two particles striking the phase detector at the same time) can nevertheless cause a cycle slip in the PLL. Since the divider is also implemented with TMR, it generates 3 feedback clocks FBA-FBC which should be identical at all time. In this way a full TMR chain is implemented from the divider to the phase detector. Thus, an SEU hit

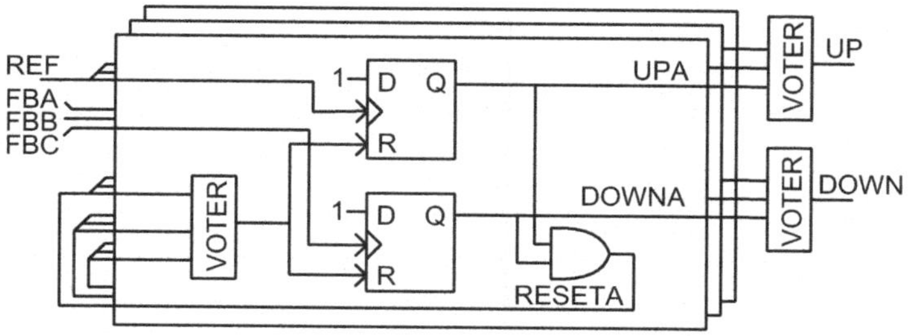

Fig. 5.12 Circuit schematics of a TMR PFD with asynchronous reset

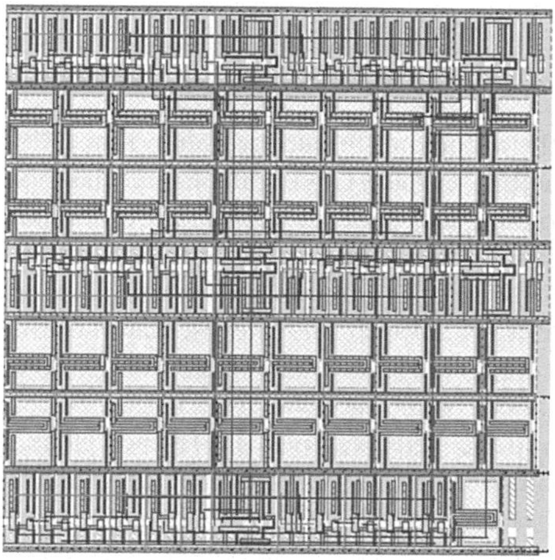

Fig. 5.13 Example layout of a TMR PFD

on a voter at the output of the divider does not generate an error in the PLL. Note that the reference signal is shared among the 3 PFDs since only one signal enters the chip. The layout of this circuit is shown in Fig. 5.13. Each individual PFD is placed in one row and is spaced to the next PFD by 2 rows. An interconnection between the rows is required by the reset signals and, at the output to vote the up and down signals.

5.7 Automatic Frequency Calibration

The discrete capacitors of the oscillator are implemented with MOMCAPs, they have a very good Q factor compared to MOSCAP varactors but suffer from large variability in different process corners. The result is that the frequency bands shift in different corners. These variations can be so large that the correct preset of the banks cannot be guaranteed in advance such that the PLL can never lock. Therefore, an automatic frequency calibration algorithm is included in this chip. During startup,

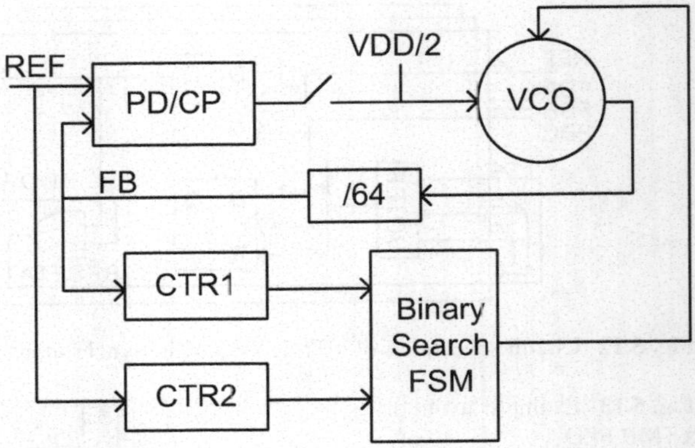

Fig. 5.14 AFC system architecture

the automatic frequency calibration (AFC) algorithm is launched and the correct setting of the VCO is calculated. The logic is synthesized to include triple modular redundancy and all registers are continuously clocked to prevent the calculated values from being lost after an SEU hit.

Figure 5.14 shows the system components of the AFC. During calibration, the PLL is stopped and the oscillator is free running with a constant voltage. Two counters are reset and start running. One counts on the reference clock while the other counts the divided oscillator. Depending on the preset value of the oscillator and the frequency of the reference clock, either the first or the second counter will timeout first. A binary search algorithm updates the preset value of the oscillator based on which timer was first completed and the counters are reset and ran again. For each bit (4 bits in this design), there has to be a count and an update of the bits. Thus, the calibration time is constant and mainly determined by the counter depth. Both counters count asynchronously to the FSM but the timeout-flags and reset events are synchronized to the FSM. The counters are designed as sequential ripple counters where each stage is triplicated for SEUs. The FSM uses traditional TMR to protect the logic against radiation effects.

5.8 Measurements

This section will provide the measurements of this circuit, not related to radiation effects.

5.8.1 VCO Characteristics

The free running oscillation frequency of the LC-tank oscillator and ring oscillator is shown in Fig. 5.15. These measurements were done using a high speed digital

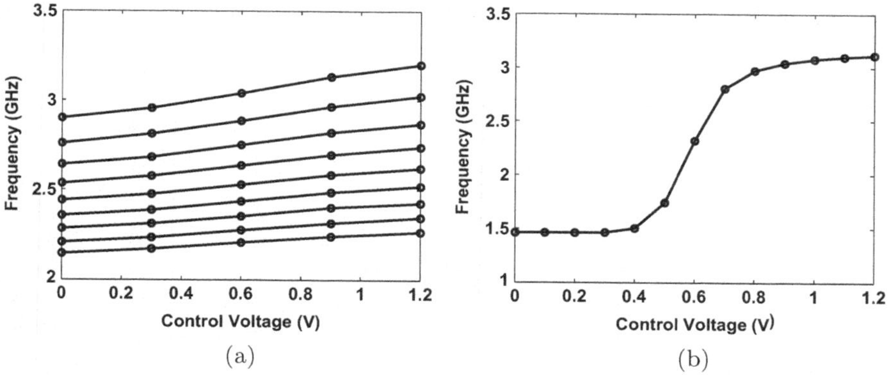

Fig. 5.15 Free running oscillation frequency. (**a**) LC-tank oscillator. (**b**) Ring oscillator

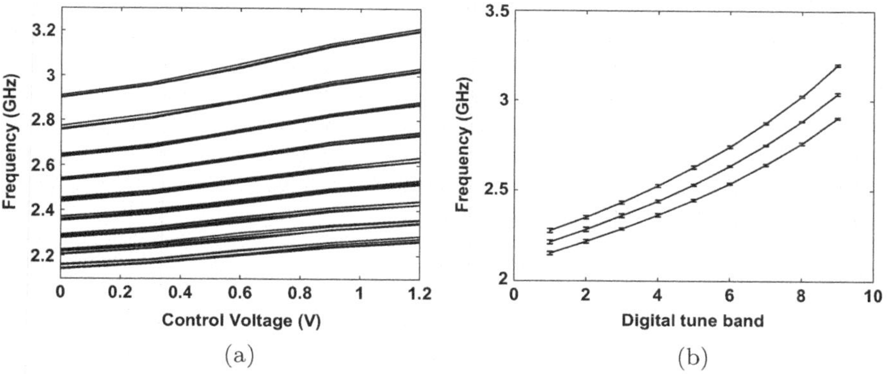

Fig. 5.16 Local process variations. (**a**) Free running LC-tank oscillation frequency for 5 different samples. (**b**) 1σ error bars for Vtune = 0, 0.6, and 1.2 V

sampling oscilloscope. The measurements show that the overlap which has been designed between different banks is sufficient for the AFC to find a solution. This ensures that the AFC will always find an optimal setting so that the charge-pump can operate in a range from 1/4 Vdd to 3/4 Vdd for a good PSRR. It is clear that the ring oscillator has much higher gain as mentioned previously. Compared to a gain of 250 MHz/V for the LC-oscillator, the ring-oscillator has a gain of 5 GHz/V. This also extends to a much wider tuning range for the ring-oscillator compared to the LC-oscillator. However, for the targeted application, the tuning range for the LC-tank is sufficient.

Five samples were measured to study the process variation between different chips from the same batch. Figure 5.16 shows an error bar plot for control voltages 0, 0.6, and 1.2 V across different digital bands in the LC-oscillator. The maximum

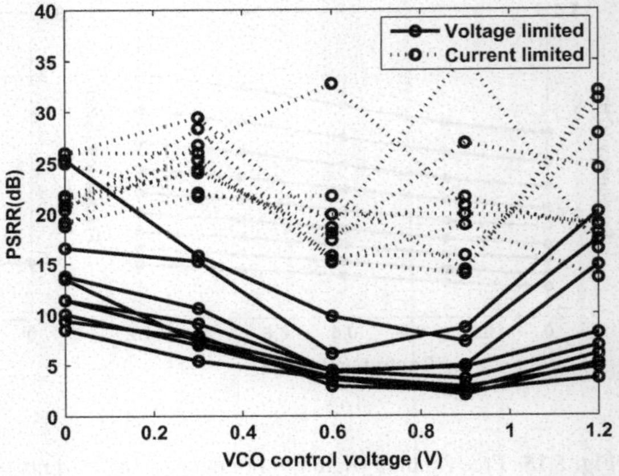

Fig. 5.17 PSRR measurements for different digitally controlled tuning bands

measured error (standard deviation) was 11 MHz which is only 0.4% of the center frequency. It can be concluded that local variations are not significant in this design.

The LC-oscillator has two modes, either voltage limited or current limited mode. In voltage limited mode, the power supply rejection mode is expected to be smaller compared to current limited mode since the output resistance of the current source, loaded with the active cross coupled pair should filter part of the supply noise. To measure this, the free running oscillation frequency was measured for different supply voltages. The PSRR was defined as

$$PSRR = \frac{2\Delta f/\Delta V_{dd}}{\langle \Delta f/\Delta V_{tune} \rangle} \tag{5.10}$$

in which $\langle \Delta f/\Delta V_{tune} \rangle$ is the average VCO gain in the band. The factor 2 originates from the voltage division ratio between the pmos and nmos cross coupled pair. The measurements of the PSRR are shown in Fig. 5.17. This plot shows the PSRR across different tuning voltages from which it is clear that the PSRR has a dependency on the tuning voltage. This can be explained since the varactors become more sensitive towards the center of the tuning range.

5.8.2 Bandwidth

The bandwidth of the PLL was measured using a reference clock with white Gaussian jitter which has a 50 ps standard deviation at the 40 MHz reference clock as is shown in Fig. 5.18. The output phase noise of the PLL was then subtracted from the phase noise which originates from the reference generator. If the reference is not completely white, then this effect will be canceled during the subtraction. Figure 5.19 shows the transfer function of the LC-oscillator PLL for two extreme

Fig. 5.18 Measurement principle of the PLL bandwidth

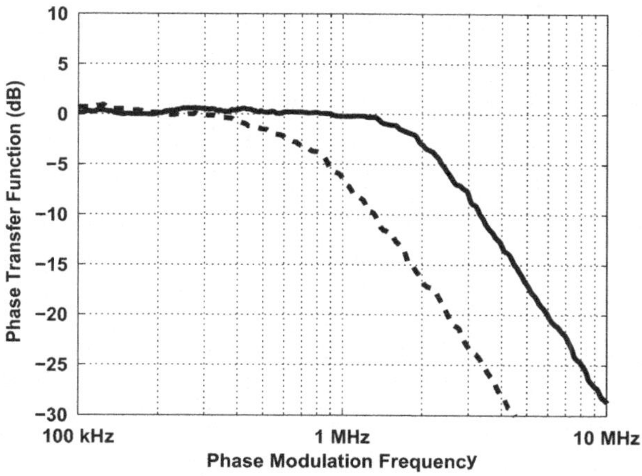

Fig. 5.19 PLL bandwidth measurement for a high and low bandwidth configuration

settings. The charge-pump current and resistor value can be digitally configured to adjust the bandwidth of the PLL with a constant damping factor. The measurements show that the −3 dB bandwidth of the circuit can be adjusted from 700 kHz up to 2 MHz.

5.8.3 *Phase Noise*

The phase noise of the LC-oscillator PLL was measured with a phase noise analyzer (R&S FSW), both in open and closed loop configuration. Figure 5.20 shows the measured phase noise for 5 different samples. The close-in phase noise is measured at −110 dBc/Hz at an operating frequency of 2.56 GHz. The reference clock was a high precision signal generator with 200 fs rms jitter. The total jitter measured, including signal generator, input receiver, PLL, and output drivers was 345 fs rms. This jitter is calculated from the phase noise measurement in an interval from 100 kHz to 100 MHz. Figure 5.20 also shows the open loop phase noise of the oscillator which has a typical noise of −124 dBc/Hz at an offset frequency of 1 MHz which yields a FOM of

$$FOM = -PN(f) + 20\log\left(\frac{f_0}{\Delta f}\right) - 10\log(P_{\mathrm{mW}}) = 185\,\mathrm{dBc/Hz}_{@1\mathrm{MHz}} \tag{5.11}$$

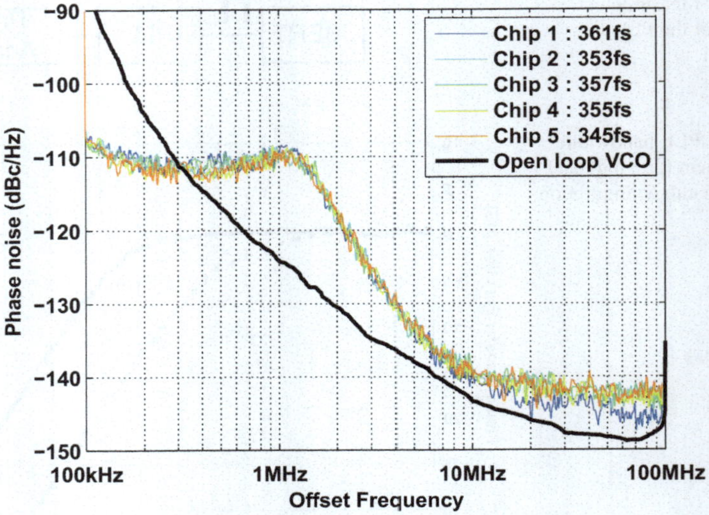

Fig. 5.20 Phase noise measurement of 5 different samples. Open loop VCO phase noise as a comparison

These measurements were done with the current source of the oscillator bypassed to the supply rails.

The core power supply consumption of the PLL is 11.7 mW and has a FOM of

$$10 \log \left(\frac{\sigma_{\tau_j}}{1s} \cdot P_{1\,\mathrm{mW}} \right) = -238.5\,\mathrm{dB} \tag{5.12}$$

A similar measurement was done using a high performance oscilloscope (Tektronix DPO 7000) which captures the output waveform of the oscillator. The internal specifications of the oscilloscope reported a timing uncertainty of 290 fs rms which is subtracted quadratically from the calculated rms jitter. Figure 5.21 shows the noise measurements referenced to an ideal reference clock. The crossing points are calculated from the time domain waveforms. The measured jitter has a standard deviation of 485 fs from which the internal jitter of the oscilloscope has been subtracted. The jitter histogram also shows a Gaussian match of the noise.

Finally, also the phase noise of the ring-oscillator PLL has been measured and is shown in Fig. 5.22. This measurement shows, as expected, that the phase noise of the ring oscillator is much higher compared to the LC-tank oscillator. The integrated phase noise from 100 kHz up to 100 MHz was measured to be 5.2 ps rms compared to 345 fs rms in the LC-tank oscillator PLL.

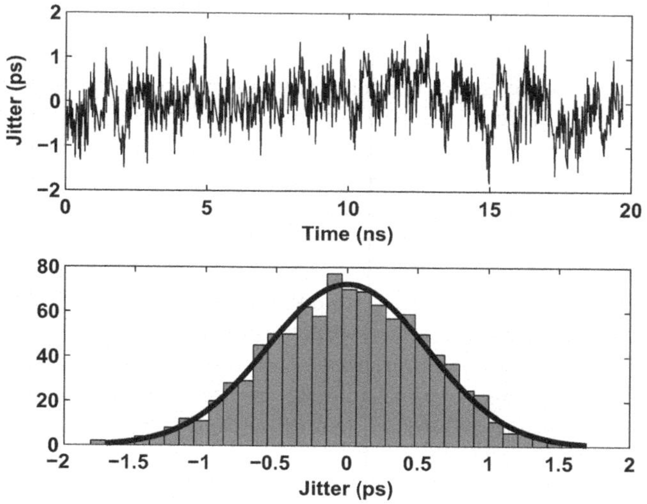

Fig. 5.21 Time domain jitter measurements using a sampling oscilloscope with a calculated jitter histogram

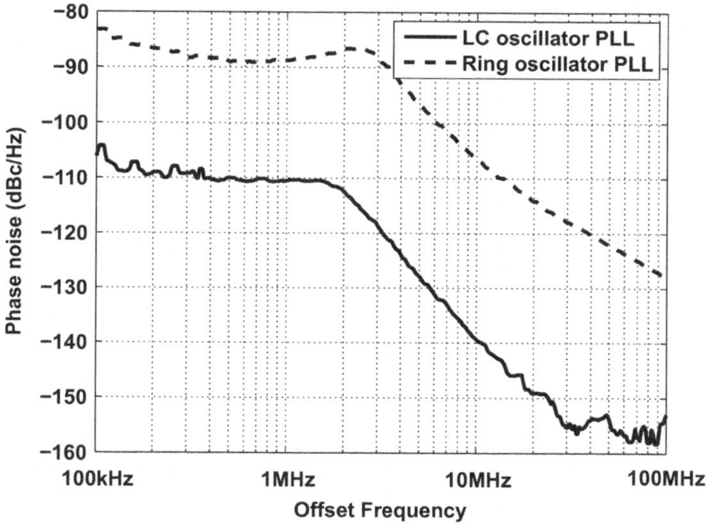

Fig. 5.22 Phase noise comparison of the LC-oscillator and Ring-oscillator PLL

5.8.4 Temperature Variations

The circuits have been tested for temperature variations to estimate the robustness of the devices. The ICs were placed in a climate chamber and were tested from −25 °C up to 125 °C. During each temperature step, the free running frequency, phase noise, and transfer function were measured. Free running LC-oscillator measurements are shown in Fig. 5.23a,b. The measurements for the ring-oscillator are shown in Fig. 5.23c. From these numbers, it can be concluded that the frequency deviation in an LC-tank oscillator is significantly smaller than the frequency deviation in a ring-oscillator. This can be explained from the fact that the resonance frequency of an LC-tank is relatively temperature independent compared to the ring-oscillator's frequency which is set by mos transistor currents. The current is temperature dependent which explains this measurement. Figure 5.23d shows the closed loop PLL transfer function for different temperatures. At higher temperatures, the

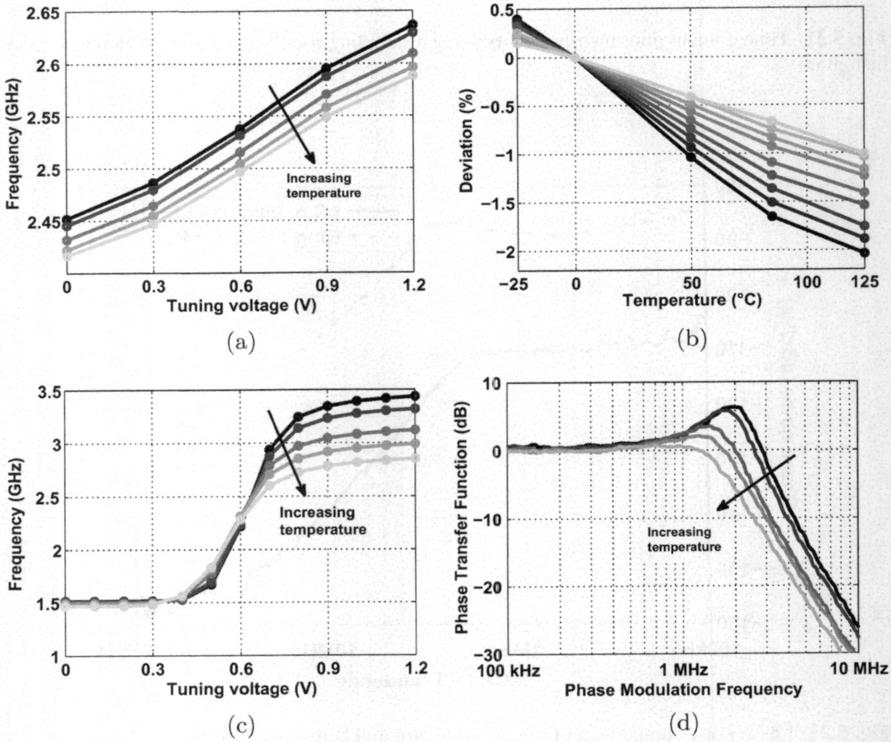

Fig. 5.23 PLL pre-radiation measurements. (**a**) LC-oscillator free running frequency. (**b**) LC-oscillator frequency deviation for different tuning bands. (**c**) Ring-oscillator free running frequency. (**d**) PLL transfer function

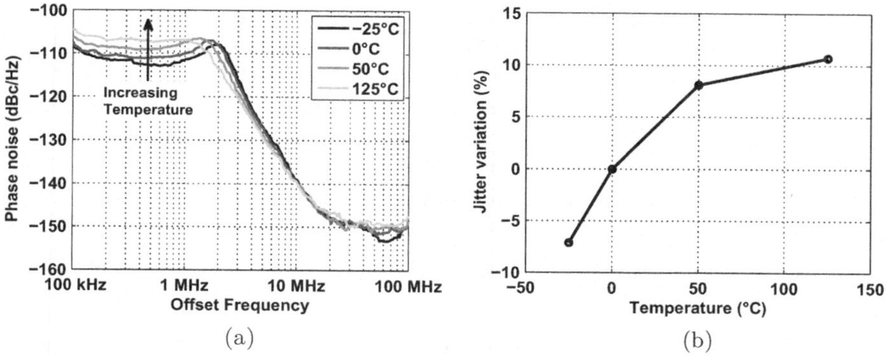

Fig. 5.24 Influence of the temperature on the phase noise. (**a**) LC-oscillator PLL phase noise from −25 to 125 °C. (**b**) Integrated jitter for different temperatures

bandwidth of the loop reduces and the damping factor increases. This effect can be explained from increasing loop filter resistance and decreasing charge-pump current.

The closed loop LC-PLL phase noise measurements are shown in Fig. 5.24a. From this measurement, it can be seen that at high temperatures, the in-band phase noise increases. Firstly, the thermal noise increases with temperature. Secondly, due to a slight reduction in the loop bandwidth and a reduction of the charge-pump current, the in-band phase noise increases. The phase noise plots can be integrated to acquire the integrated jitter in the PLL which is shown in Fig. 5.24b. This calculation shows the variation of the integrated jitter in the LC-oscillator PLL across the temperature range. In total, a variation across its range of 17% is observed which lies within the noise/jitter specifications taking into account the nominal integrated jitter.

5.8.5 Layout

A chip photograph of the devices is shown in Fig. 5.25a,b. From this picture it can be seen that the LC-tank oscillator occupies a significant amount of area due to the inductor and capacitors. In the ring-oscillator PLL however, most area is occupied by the loop filter since the ring-oscillator itself is small compared to the capacitors.

A comparison to relevant state-of-the-art circuits is shown in Table 5.3. This comparison shows that the presented PLL is competing with current circuits in terms of performance with an additional radiation hardness which was not reported in the other works.

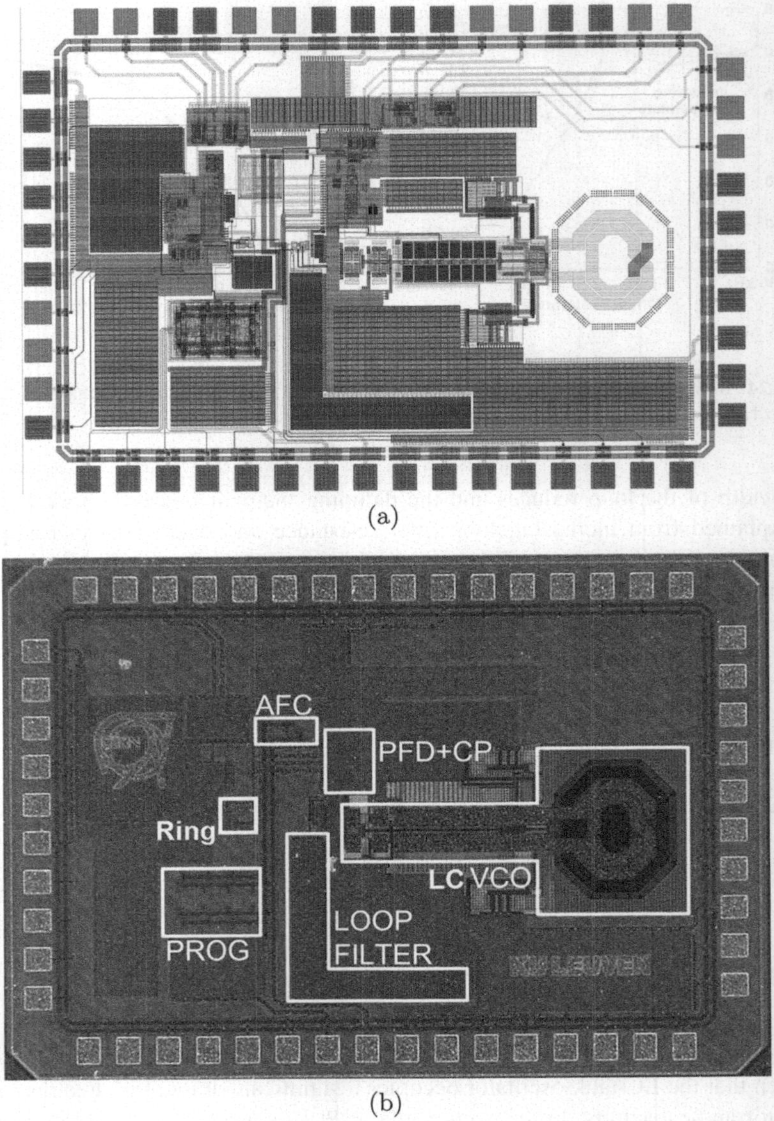

Fig. 5.25 Chip photograph

Table 5.3 Performance summary and comparison

	This work	[147]	[148]	[149]
Technology	65 nm	180 nm	180 nm	65 nm
Freq. (GHz)	2.2–3.2	2.2–2.4	2.5–3.2	2.9–4.0
BW (MHz)	0.7–2	0.5	0.2	0.3
RMS jitter (fs)	345	266	455	560
Power (mW)	11.7	17.3	48	4.5
FOM (dB)	−238.5	−239.1	−230	−238.5
Rad. tolerant	Yes	No	No	No

5.9 Conclusion

This chapter introduced a prototype chip which will be used in high-resolution multi-channel TDCs as a clock source for the timing generator. A practical background on radiation effects in charge-pump PLLs was discussed with the major effects on the charge-pump/loop filter, VCO, PFD, and divider, followed by the discussions of the circuit. The circuit includes two charge-pump PLLs with a ring-oscillator and LC-oscillator which are processed in a 65 nm 1.2 V CMOS technology. The ring-oscillator has a tuning range from 1.5 GHz up to 3.2 GHz and the measured closed loop PLL jitter was 5.2 ps rms. However, temperature experiments from −25 to 125 °C showed a significant dependency on the free running oscillation frequency. The low-noise LC-tank oscillator can be tuned from 2.2 to 3.2 GHz with a typical phase noise of −124 dBc/Hz @ 1 MHz and has a FOM of 185 dBc/Hz. The PLL has a configurable bandwidth from 0.7 to 2 MHz. The measured jitter of the full circuit is only 345 fs rms at a power consumption of 11.7 mW resulting in a FOM of −238.5 dB.

Radiation experiments on these two circuits, both TID and Single-Events, are discussed in the next chapter.

Table 5.5 [illegible] and comparison

[illegible]	[illegible]	[illegible]

5.9 Conclusion

This chapter introduced a prototype [illegible] multi-channel TDC [illegible] [illegible] the [illegible] the dimensions of the [illegible] [illegible] [illegible] The [illegible] [illegible] [illegible] FDM of 125 [illegible] [illegible]

[illegible] discussed in the next chapter.

Chapter 6
Radiation Experiments on CMOS PLLs

6.1 Introduction

In this chapter, the radiation experiments on the LC-oscillator and ring oscillator PLL are discussed. First, a discussion is made on the TID effects which are measured using X-rays. In total, two chips were irradiated to compare their results. Next, the single-event sensitivity of the circuits is discussed through measurements done with heavy ions and laser experiments. Finally, a brief discussion is done on the practical aspects of the setup.

6.2 TID Effects

The circuits were irradiated with X-rays with a 40 keV, 50 mA W-tube which results in a (calibrated) dose-rate of 9 Mrad/h (SiO_2) at the X-ray facility at CERN [150]. Using a 2 cm beam-diameter, the chip could be uniformly irradiated. Two samples were tested. The first one was irradiated up to 200 Mrad and annealed for 24 h at room temperature. In the second step, annealing at 100 °C was performed and measured after 4 and 24 h. The annealing was performed with the circuit biased in normal operating conditions, which provides the most realistic condition. The second sample was irradiated up to 600 Mrad and measured after 24 h room temperature annealing and 24 h annealing at 100 °C. During irradiation, the free running oscillation frequency of the oscillators was measured at various points in time to study the effects on the oscillator. Also, the closed loop noise measurements were done at those times.

J. Prinzie et al., *Radiation Hardened CMOS Integrated Circuits for Time-Based Signal Processing*, Analog Circuits and Signal Processing,
https://doi.org/10.1007/978-3-319-78616-2_6

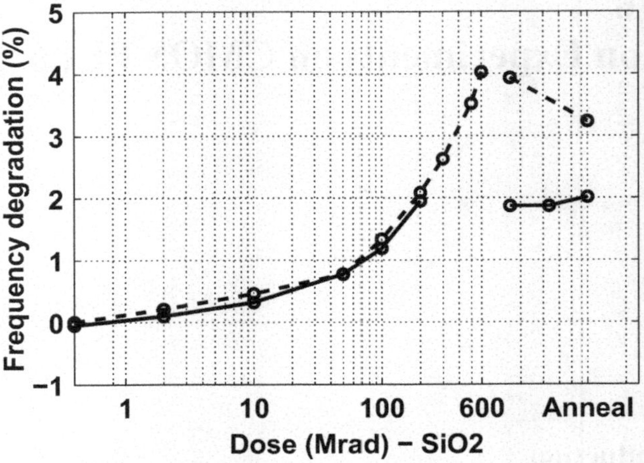

Fig. 6.1 Free running LC-oscillator frequency change for 2 different samples irradiated up to 200 and 600 Mrad (SiO_2), respectively, with a variation of only 2% to 4%

6.2.1 LC-Oscillator

The frequency of the LC-oscillator during irradiation is shown in Fig. 6.1. The irradiation of the LC-oscillator up to 600 Mrad did not show a variation in free running oscillation frequency of more than 5%. Note that only one tuning band (four capacitors switched to the LC tank) of the tuning curve of the oscillator is shown here. Also, annealing does not significantly change the behavior of the circuit. During irradiation, the PLL was operational and in lock. At different doses, the irradiation was temporarily stopped and the open loop characteristics were measured. The measurement shows two chips, one which was irradiated up to 200 Mrad while the other was irradiated up to 600 Mrad. The measurements show a good match between both experiments. Figure 6.2 shows a summary of the frequency for different digital tuning bands.

To understand the increase in frequency, the equivalent capacitance is calculated as is shown in Fig. 6.3a. The equivalent capacitor value of the LC tank is calculated from the measured oscillation frequency, assuming the inductor value remains constant ($C = 1/4\pi^2 L f^2$).

This small frequency variation originates from the change in transconductance due to TID effects on the cross coupled transistors in the oscillator which compensate for the losses in the tank. These transistors have a 120 nm, gate length. Figure 6.3b shows a simulation result of the free running oscillator obtained with a PSS (Periodic Steady State) simulation. The simulation was done by adding an ideal positive transconductance block (loss) parallel to the tank to simulate the reduced transconductance of the -Gm circuit. The reduction of transconductance of the active block results in a smaller oscillation amplitude and a reduction of the common mode

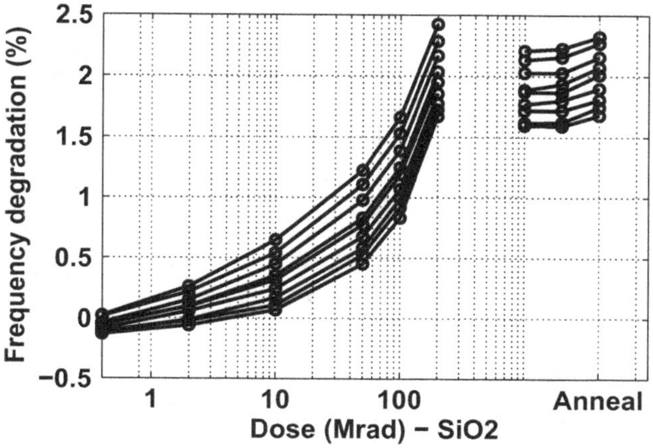

Fig. 6.2 Free running LC-oscillator frequency change for all different digital tuning bands in the VCO

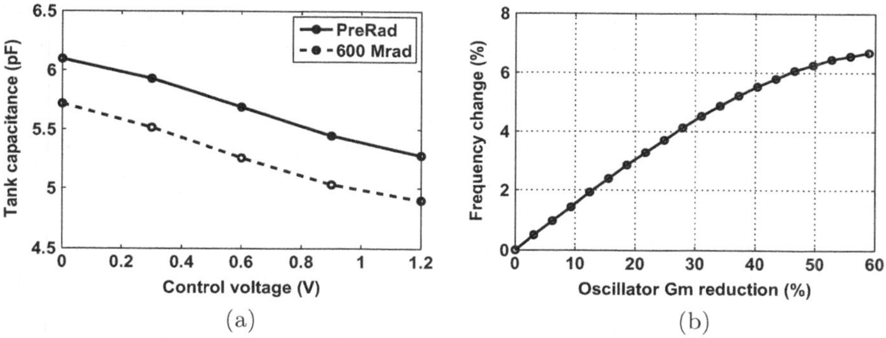

Fig. 6.3 TID effects on the LC-oscillator. (**a**) LC tank capacitance before and after irradiation up to 600 Mrad with all discrete capacitor banks enabled to the tank. (**b**) Free running LC-oscillator frequency with decreasing transconductance of the oscillator's transistors. The transconductance is decreased down to 40% of its original value

voltage of the output of the oscillator. The amplitude reduction results in a different cycle through the non-linear varactor C-V curve that yields a smaller capacitance and thus an increasing frequency. Also, the smaller DC common mode voltage results in an increase in frequency for the same control voltage due to a changed biasing point of the varactor.

Typically, the nwell-MOS varactors (Fig. 6.20), which work in depletion- or accumulation mode, show an increase in capacitance with accumulated doses due to negative induced charges in the silicon that reduce the depletion charge and increase the capacitance [151, 152]. This effect compensates for the frequency increase due to the variation in transconductance but the overall variation shows a slight increase in frequency. A likely problem to occur in the LC-oscillator at

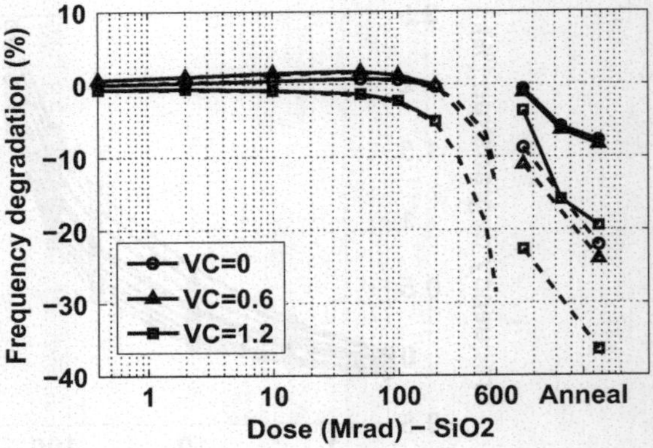

Fig. 6.4 Free running ring oscillator frequency change. Measurements done for control voltage of 0, 0.6, and 1.2 V. The solid lines represent the first sample that is irradiated up to 200 Mrad. The dashed lines represent the sample that is irradiated up to 600 Mrad. Both samples overlap the first 200 Mrad

higher doses will be a failure of oscillation when the transconductance of the cell is no longer able to compensate the LC-tank losses to sustain oscillation. The point at which the oscillation starts failing will depend on the corner case of the circuit. The oscillator has been designed to have a negative transconductance which is 3 times smaller than the losses in the tank, in typical case. In the worst case corner, the available transconductance of the active transistors will be reduced (simulated to be 2 the minimal required G_m) such that the oscillator will fail at an earlier stage. During irradiation, the circuit, which was processed in a typical corner, did not fail after 600 Mrad though the actual transconductance could not be measured since the oscillator is buffered on chip.

6.2.2 Ring Oscillator

The same measurement of the free running oscillation frequency was done for the ring oscillator which is shown in Fig. 6.4. A significant change in frequency is observed from 300 Mrad of 5% to 10% and nearly 30% of variation is seen after 600 Mrad. Compared to the LC-oscillator, a large annealing effect is observed in this circuit. According to [153], this is due to the short-channel devices in the cells for which the radiation effects are triggered during high-temperature annealing. The large variation in tuning characteristics caused the PLL to be unable to lock after being irradiated to 600 Mrad since the oscillator could no longer run at 2.56 GHz. The main cause of radiation damage in the ring oscillator lies in the dependency of the oscillation frequency on the drive current of the transistors. Since this current changes significantly, so does the frequency tuning curve. The period of the oscillator is proportional to the delay of the cells which is directly related to the transconductance of the transistors. This is not the case with LC-oscillators making them more resistant against TID effects.

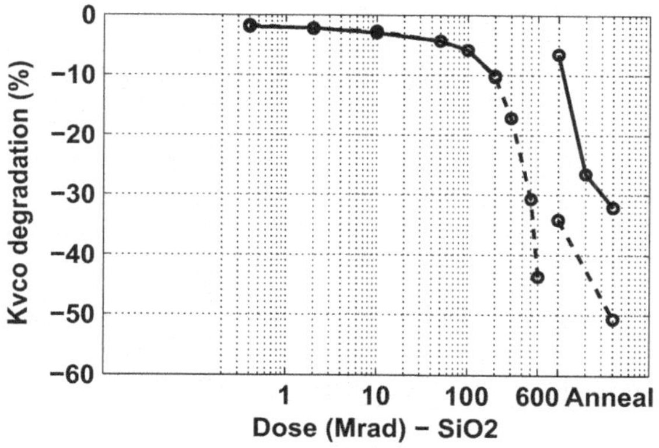

Fig. 6.5 K_{vco} of a ring oscillator irradiated up to 200 Mrad (solid) and 600 Mrad (dashed)

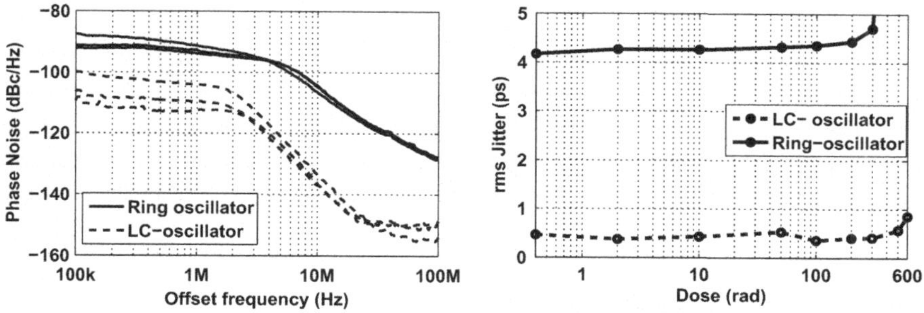

(a) (b)

Fig. 6.6 (**a**) Phase noise measurements during irradiation up to 600 Mrad. (**b**) Integrated jitter up to 600 Mrad

Another interesting graph is shown in Fig. 6.5 which shows a reduction of the VCO gain (K_{vco}) of the ring oscillator. The result is that the damping factor of the PLL reduces if the loop gain drops. From these measurements it is clear that the gain drops by more than 40% which causes the loop to be less stable.

6.2.3 PLL Phase Noise

During the irradiation experiment, the phase noise of the closed loop PLLs was also measured to monitor the jitter of the circuit at different doses. Figure 6.6a shows the closed-loop noise measurements as a function of the total dose for both circuits. In both cases, the overall noise is relatively constant. At higher doses the jitter of the LC tank increases, mainly due to changes in bandwidth (charge-pump current) [154]. The jitter of the ring oscillator increases significantly at a high dose. At a dose above 300 Mrad, the oscillator does not operate properly at 2.56 GHz due to

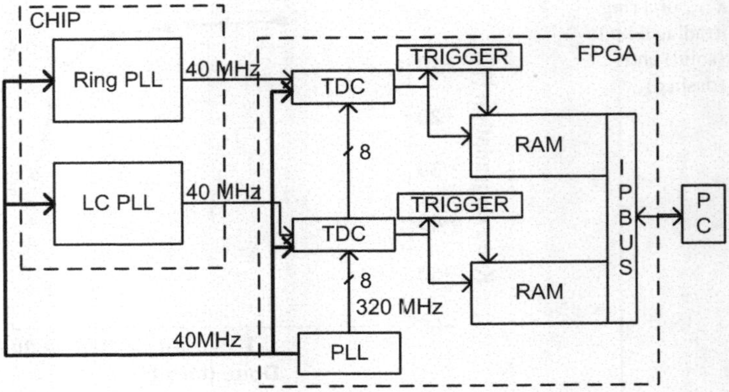

Fig. 6.7 SEU measurement setup with TDCs embedded in the FPGA

the reduced speed. A large increase of noise is observed after annealing as well. When the circuit degraded this much, the settling point of the VCO control voltage is nearly at the edge of the tuning range which results in a significant drop of the loop gain (less damping and lower bandwidth) which dramatically increases the noise.

6.3 Single-Event Effects

Single-Event Effects were studied in both PLLs to understand the effects of single charged particles on the circuits. In total, two experiments were done. The first experiment was done using heavy-ions at the Cyclotron facility at UCL, Belgium. During this test, both chips are uniformly irradiated with charged particles and statistics were recorded from the upsets. The main disadvantage of this experiment is that it can be hard to find the precise location from where the error in the circuit originates. To understand the effects in more detail, a Two-Photon Absorption (TPA) laser experiment was done.

6.3.1 Experimental Setup

When the PLLs are irradiated, phase errors need to be measured. It is interesting, not only to detect an error but also have some information on the phase signal itself. For this, a custom, two-channel TDC has been designed that is embedded in a Xilinx Spartan 6E FPGA as is shown in Fig. 6.7. This dedicated setup allows the experiment to be easily done without expensive measurement equipment. Two identical TDC channels are implemented with a timing bin-size of 390 ps. The TDCs have a sample-rate of 40 MHz which is the speed of the reference clock and it

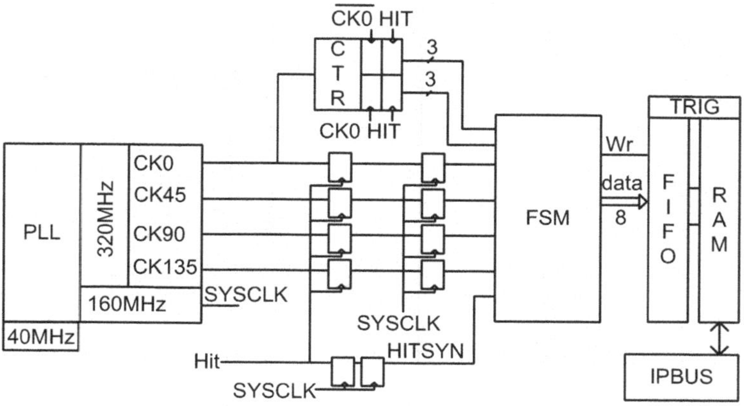

Fig. 6.8 SEU measurement setup with TDCs embedded in the FPGA

triggers on phase jumps of more than 390 ps in the TDC channels. Once a trigger occurs, 1024 successive data samples (25.6 μs) of the TDC are stored in a RAM and sent to a PC over an Ethernet connection. In this way, transients in the PLL can be studied after the experiments. The chip boards were located in the heavy-ion or laser beam while the FPGA setup was located outside the chamber, such that no SEU problems can occur in the FPGA.

A detailed architecture of the TDC is shown in Fig. 6.8. A PLL in the FPGA is used to generate different clocks in the circuit. Firstly, a 40 MHz clock is generated which is used to drive the reference clock of the chips, secondly, a 160 MHz clock is generated which is the system clock in the digital core of the TDC decoding and interface logic. Thirdly, four 320 MHz clocks (0°, 45°, 90°, 135°) are synthesized which are used to measure the time differences. These four phases can be used to interpolate the 320 MHz clock 8 times (eight phases) resulting in a bin size of 390 ps.

The hit-signal in this architecture is the output clock from the PLL chip and samples the 320 MHz phases, similar to a delay-line TDC from Chap. 2. This sampled data is synchronized and being decoded by the logic to measure the time difference between the hit-signal and the 320 MHz clock. However, the 320 MHz clock runs 8 times faster than the 40 MHz clock such that a 3-bit counter is required to cover the full range of 25 ns. Since the hit-signal is asynchronous (for example, during a transient), it may sample the counter while it is toggling. To prevent metastability errors, two counters are used which run on the rising and falling edge of the 320 MHz clock. The correct value is later selected by the logic, depending on the output of the fine-time measurement using the eight phases. This is similar to the extended counter, used in TDCs as discussed in Chap. 4. The core of the TDC periodically measures the time difference between the hit-clock (chip output) and the reference clock. This data is pushed in a FIFO and is interpreted by a trigger logic module. Once triggered, the data is forwarded to a dual-port RAM memory which can be read through Gbit Ethernet. A python application on a computer/laptop can store this data to a hard drive.

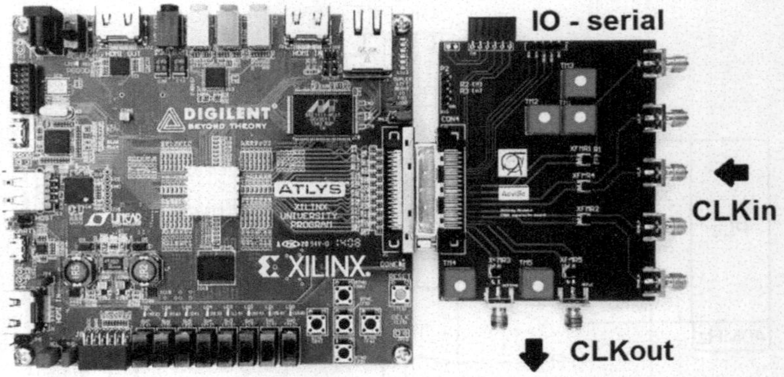

Fig. 6.9 FPGA with expansion board to connect the circuit to the FPGA board

The bottleneck in this system is the interface to the computer/storage. Currently, the computer polls for new data and reads the RAM block. This polling takes some time. To overcome the read latency when the block is read, two RAM blocks are used in each channel such that new data can already be pushed in a slave memory while the other RAM is read by the computer. Each cycle, the active/slave RAM is swapped but this is transparent to the TDC or computer through a small digital memory manager. Still, if a few errors occur very close to each other, the RAMs may still be full and the measurement (1024 phase measurements) is dropped. In the FPGA, a counter is located which measures the number of errors. Even if the phase measurements are not stored (in the case when the data rate is too high), the counters will still be incremented such that the cross-section can be accurately estimated. This was implemented in advance to ensure that no errors were missed. However, from the practical experiments, no lost packets were observed.

A picture of the FPGA test setup is shown in Fig. 6.9. A commercial FPGA board was used with a custom expansion board that has baluns to interface from the 1.2 V chip supply to the 3.3 V FPGA power domain using differential signals. Figure 6.10 shows a picture of the small PCBs that contains the prototype chips. This board has two differential output pairs which are the high-speed 2.56 GHz output clocks of the PLLs. The 40 MHz reference and output clocks are distributed to the main board through three connectors which also supply power and control signals. The main board is shown in Fig. 6.11. On this PCB, the power (1.2 and 3.3 V) is generated. In this way, it is easy to test different chips with the same test setup (Figs. 6.10 and 6.11).

6.4 Heavy-Ion SEU Tests

SEU tests are done using cyclotron accelerated heavy-ions which had a Linear Energy Transfer (LET) of 3.3 to 69.2 MeV $\cdot$ cm^2/mg (Table 6.1). When the chip was irradiated, the SEU count was monitored as shown in Fig. 6.13. The cross-sections

Fig. 6.10 PCB substrate that holds the prototype chip and can be connected to the main board

Fig. 6.11 Main PCB that holds the small chip PCBs and supplies power and reference clocks to the chip

of both PLLs, according to the number of upsets, are shown in Fig. 6.12. In this context, an upset is defined as a phase jump that was captured by the FPGA. For each test, a particle fluence of $8 \cdot 10^6/\text{cm}^2$ was used with an average flux of $11 \cdot 10^3/\text{s} \cdot \text{cm}^2$. These measurements show a significantly higher sensitivity to SEUs on the LC oscillator compared to the ring oscillator PLL. One difference between both PLLs also lies within the charge pump. The ring oscillator based PLL has

Table 6.1 Heavy ions used to test the single-event effects

	LET (MeV · cm^2/mg)	Range Si (μm)	Energy (MeV)
$^{124}Xe^{35+}$	62.5	73.1	995
$^{84}Kr^{25+}$	32.4	94.2	769
$^{58}Ni^{18+}$	20.4	100.5	582
$^{40}Ar^{12+}$	10	120.5	379
$^{22}Ne^{7+}$	3.3	202.0	238

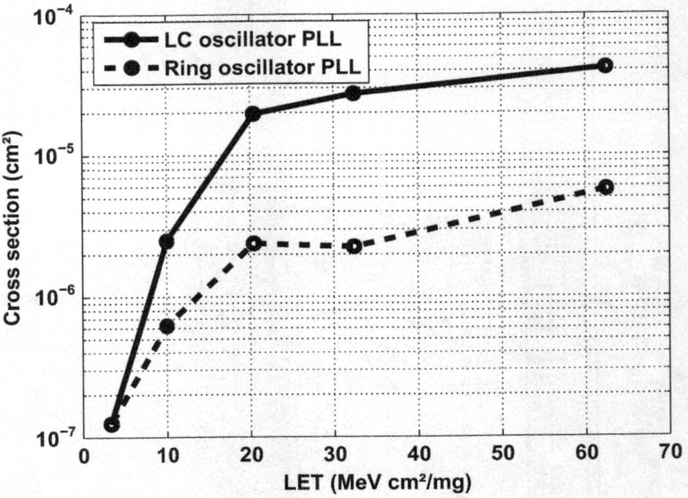

Fig. 6.12 SEU cross-section from heavy ion tests

lower charge-pump current which would result in a higher sensitivity, but this is not observed. Since both PLLs have identical components (PFD, divider, IO), the main cause of the SEU difference lies in the oscillator itself. No phase-events were observed below an LET of 10 MeV · cm^2/mg.

The captured data by the TDCs for both channels is shown in Fig. 6.14. Channel 1 and Channel 2 correspond to the LC-oscillator and ring oscillator PLL for different ions. The first observation is that the time constants of the transients are significantly different in both PLLs. However, the bandwidths are measured to be very similar. The duration of the transients in the LC-oscillator PLL indicates that there is a loss of lock and not just a phase jump that has to be compensated for. The measurements of the LC-oscillator PLL show that most phase jumps are in the positive direction. This can be confirmed if the histogram is shown (Fig. 6.15) of the maximum error that is registered by the measurement setup. In this setup, a positive phase error means that the phase of the PLL is lagging to the phase of the reference clock. Secondly, in contrast to the transients in the ring oscillator, the response of the LC-oscillator does not show a real phase jump. There is a more gradual increase in phase observed that is spread over multiple clock cycles. In the case of a phase

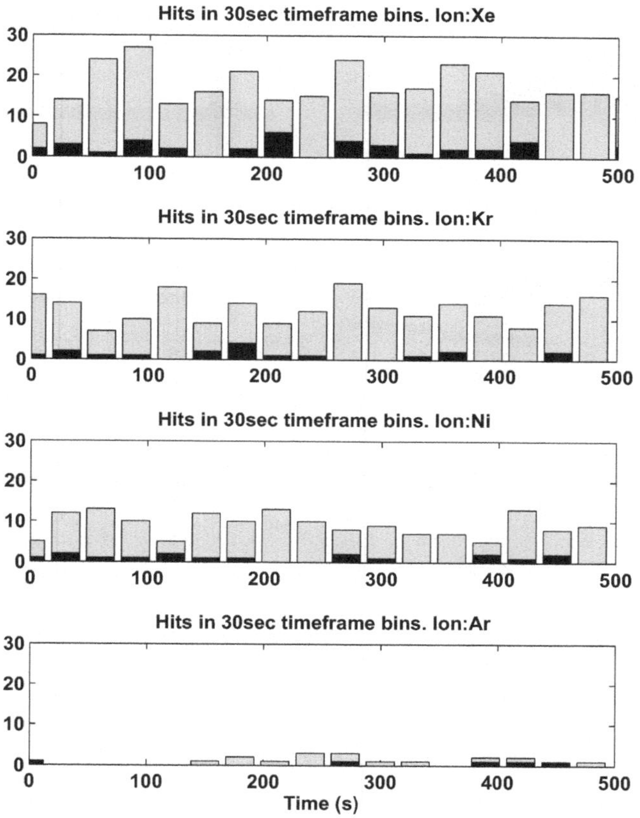

Fig. 6.13 Number of SEUs which are recorded in a 30 s time window. The grey bars indicate the errors in the LC-oscillator PLL. The black bars indicate the ring oscillator PLL

jump, the phase change would be a discrete jump in one cycle. This indicates that the event causing the phase deviations is a frequency step rather than a phase step. The frequency deviation is compensated for by the PLL. As will be discussed in Sect. 6.5, the contributor for this effect is the tuning varactor of the oscillator.

6.5 Two-Photon Laser Absorption Experiments

To gain more in-depth knowledge about the sensitive blocks of the PLL, laser tests have been done with a Two-Photon Absorption (TPA) laser. With this technique, charges can be injected in the chip at specific points with micrometer resolution. In this way it is possible to excite individual circuit blocks to identify sensitive

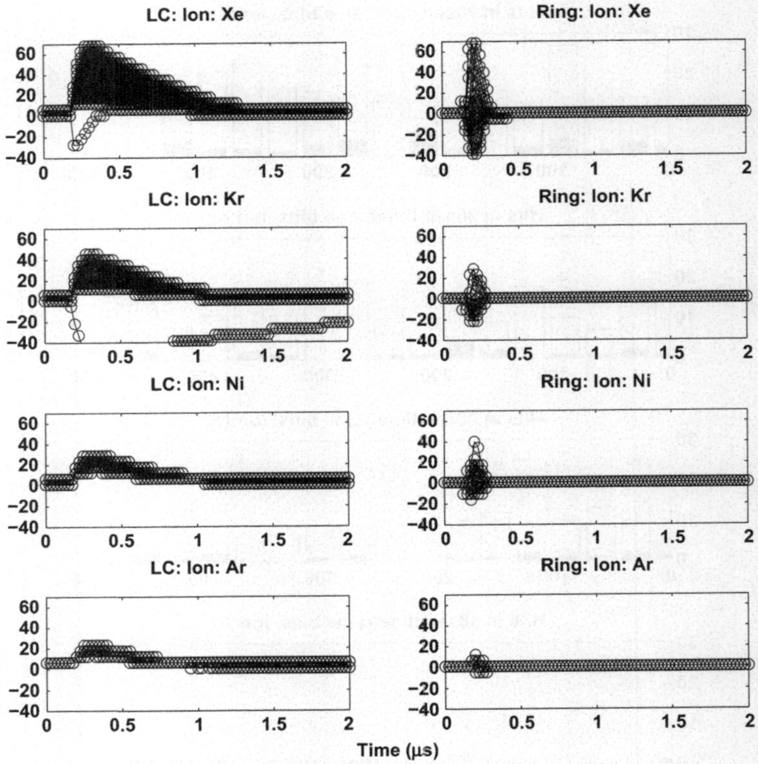

Fig. 6.14 Recorded phase errors during the experiment for both circuits and different ions

nodes [155, 156]. The laser was configured with a pulse duration of 450 fs and a 1150 nm wavelength. Unless otherwise noted, a pulse energy of 600 pJ was used at a repetition rate of 100 Hz.

A TPA laser experiment works as follows. A short pulsed laser beam is focused with its focal point in the substrate of the silicon chip (just below the active area). The wavelength of the laser is too long (the photons do not have sufficient energy) to generate electron-hole pairs in the semiconductor. However, at the location where the laser focuses, the volumetric intensity is so high that a non-linear effect, called Two-Photon Absorption occurs. What happens is that the probability becomes significant that two photons, together, generate an electron-hole pair as is shown in Fig. 6.16. This probability is proportional to the square of the intensity. It is negligible at places with low intensity but becomes dominant at high intensities. The advantage of this approach is that the laser intensity is not attenuated as it travels through the substrate towards the focal point.

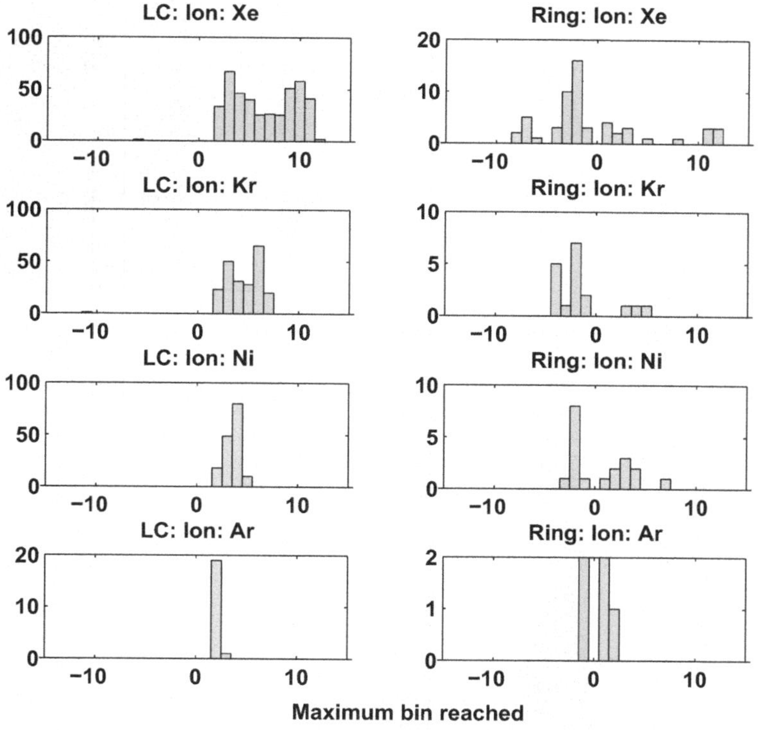

Fig. 6.15 Histogram of the maximum phase deviation bin

Fig. 6.16 Single photon and two-photon absorption (TPA)

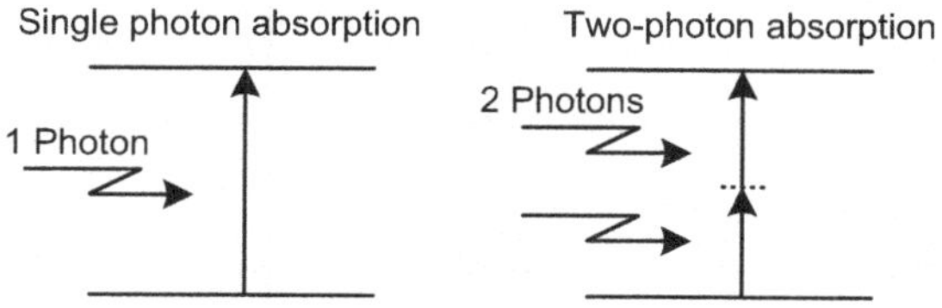

6.5.1 LC-Oscillator

All active circuit area of the LC-oscillator was scanned with the laser but only the varactors showed a response to the charge injection. The -GM cells, bias or capacitor bank switches did not show any sensitivity. When injecting charge on the varactor, similar phase deviations were measured like the heavy-ion tests with only positive phase deviations. Figure 6.17 shows the sensitivity of both varactors that was acquired during the laser tests. Figure 6.18 shows a comparison of the laser tests and the data acquired from the heavy ion tests. Both waveforms show a high similarity, especially in terms of recovery time and direction. Different laser

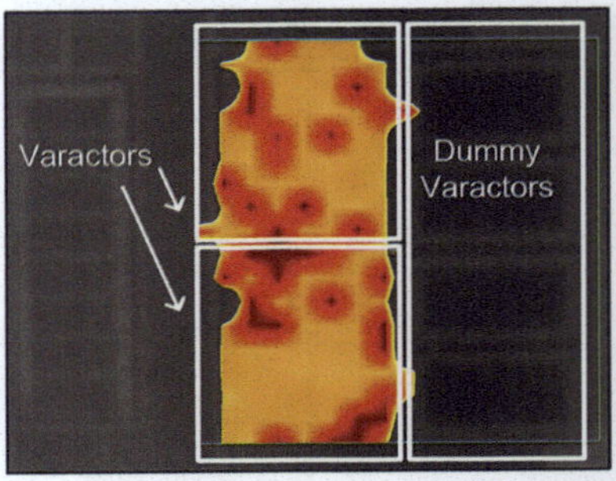

Fig. 6.17 Laser pulse injection scan on the varactor of the LC-oscillator

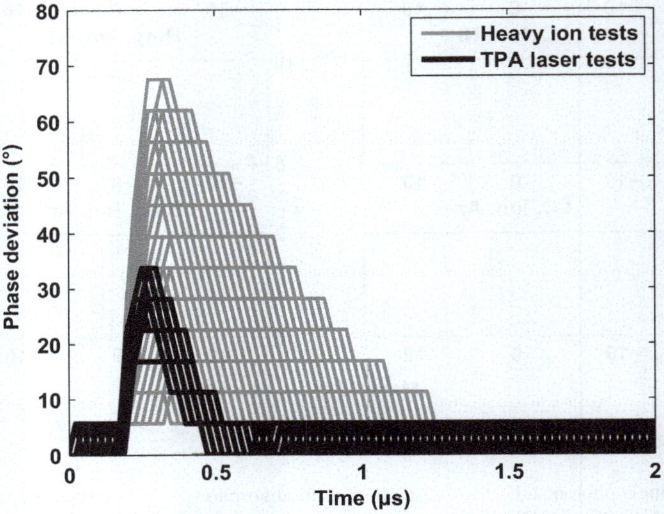

Fig. 6.18 Comparison between heavy ions and TPA laser on the varactors with 2 nJ laser energy

energies also show different amplitudes (Fig. 6.19a, b). The area of the varactor has similar cross-section as measured during the heavy-ion tests that results in the high sensitivity. While the experimental data from the laser tests show a relatively constant amplitude, the SEU data doesn't. This is due to the constant laser power that is accurately controlled and the deposited energy being constant over different injection pulses. This is not the case with heavy ion tests in which the particle energy and deposited energy are statistically distributed. The increasing amplitude with increasing laser energy shows that this is an analog event that is occurring in the circuit. The large cross-section of the varactor and phase direction can be explained by looking at the cross-section of the nMOSCAP varactor in this CMOS technology as is shown in Fig. 6.20.

Referring to Fig. 6.21, the varactors' gates (terminal A, as indicated in Fig. 6.20) are connected to the LC-tank nodes that are oscillating at 2.56 GHz while the nwell (terminals B) is connected to the control voltage of the oscillator, in this case the

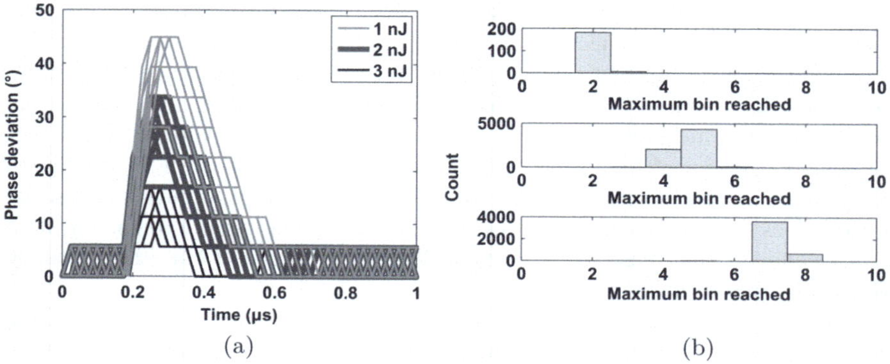

Fig. 6.19 TPA comparison of different laser energies. (**a**) Phase deviations for 3 different laser energies. (**b**) Histograms for laser energies of 1, 2 and 3 nJ

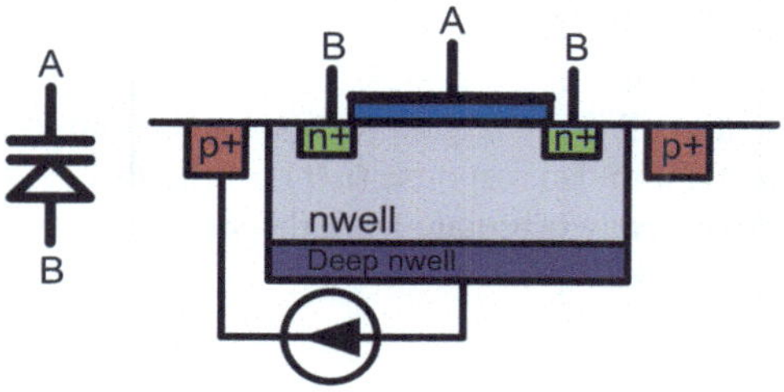

Fig. 6.20 Cross-section of the nmosCAP varactor

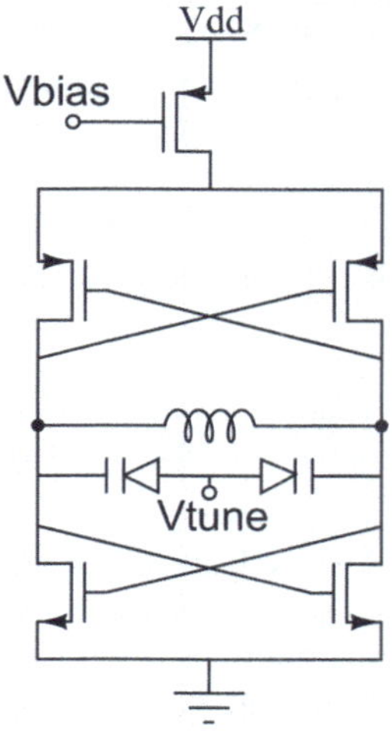

Fig. 6.21 LC-oscillator circuit

charge-pump and loop capacitor of the PLL. Below the nwell a deep nwell is buried and the p-substrate is biased to ground. While only a small portion of the varactor actually acts as a capacitor (the area below the gate), the entire nwell acts as collection junction for SEUs. Any charge injected by particles or the laser source can be separated in the depletion layer between the (deep) nwell and the p-substrate. In this way a current is injected that flows from the nwell to the substrate. In this PLL circuit, this means that a pulsed current occurs that always drains charge from the loop filter's capacitor resulting in a reduction of the voltage that controls the oscillator. Since the gain of the VCO is positive (lower control voltage gives lower frequency), the injected currents will cause a temporary reduction in oscillation frequency that results in a phase lag, like observed in the experiments. Thus, the charge does not generate an instantaneous phase jump but rather a frequency jump that accumulates the phase error over time. The PLL will recover once the frequency deviation is removed.

The sensitivity resulting from this mechanism can be prevented by avoiding the nwell connection to the control voltage of the charge pump. Simply reversing the varactor would remove the nwell from the loop filter but adds this to the oscillator nodes. In this way frequency steps are avoided but phase steps are now encountered on the oscillator. A better solution is to omit the nwell and bias it to the ground node while using a capacitive coupling to the oscillator. This solution is discussed in the next chapter.

Recall that charge injections at the tuning voltage of the VCO have a shape which is the impulse response ($h(t)$) of the PLL. A charge Q injected at the loop filter (the charge may originate from the varactors) will cause a phase error function

$$\phi(t) = \frac{2\pi}{I_{cp}} h_{cl}(t) Q \tag{6.1}$$

in which $h_{cl}(t)$ is the closed loop impulse response of the PLL. This impulse response can be obtained from spice/functional verilog simulations or system calculations based on the loop dynamics of the PLL. Figure 6.22 shows a comparison between the calculations of the PLL dynamics and the laser experiments. The charge Q is fitted to the calculations since this cannot be predicted without detailed knowledge of the technology which is not disclosed by the manufacturer.

6.5.2 *Ring Oscillator*

Results from the scanning over the ring oscillator's active area show only few sensitive hits on the core of the oscillator (delay cells). The sensitive nodes on this circuit are the oscillation nodes of each stage. The largest cross-section that is measured on the ring oscillator is the bias circuit that generates two voltages (V_{bp} and V_{bn}) that control the delay of the ring oscillator. Figure 6.23 shows the result from a scan of the biasing circuit. A charged particle will be able to disturb the bias voltage of the nmos Maneatis load [146] of the oscillator. In this way a change in frequency is induced. Both positive and negative phase deviations were observed since the node V_{bp} is connected to both nmos and pmos transistors.

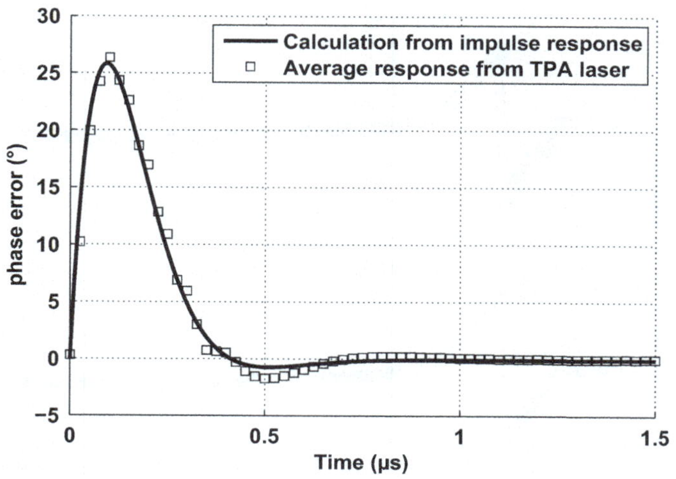

Fig. 6.22 Comparison between impulse response calculations and the real response from TPA experiments. The squares are the recorded points from the FPGA and average over all observed samples

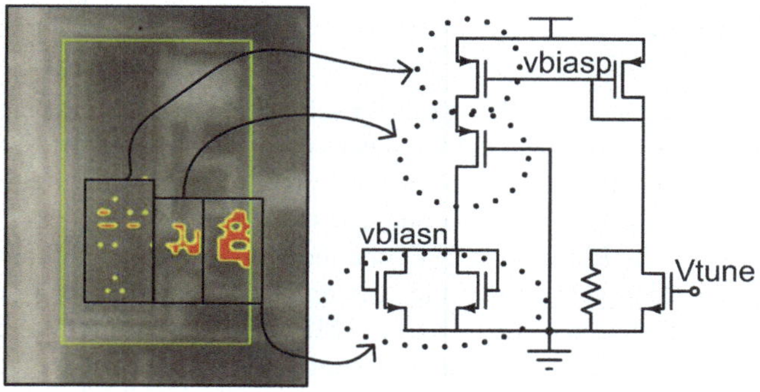

Fig. 6.23 SEU sensitivity of the biasing circuit from the ring oscillator

To reduce the sensitivity of the ring oscillator, a decoupling capacitor can be added to the biasing circuit to stabilize the biasing voltage. In lock, the bias circuit remains constant but during acquisition the bias depends on the control voltage. Therefore, the capacitance cannot be made infinitely large for the stability of the PLL. However, the core of the ring oscillator shows little space for improvement. An increase in the current of the cell will reduce the phase deviation but, since the size of the transistors will increase proportionally, the cross-section will increase as well.

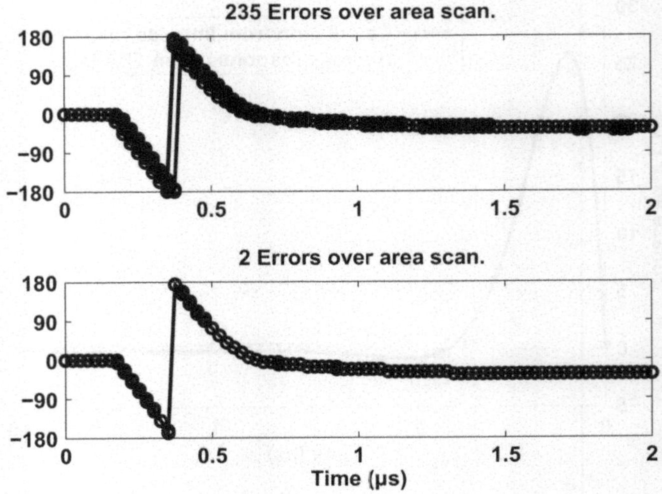

Fig. 6.24 Cycle slips for a non-TMR PFD (top) and TMR PFD (bottom)

6.5.3 *TMR PFD*

The PLL test circuit includes the option to select either the TMR phase detector or an identical single (non TMR) phase detector. In this way it is possible to fairly compare the triplicated version to a standard phase detector circuit. A laser scan is performed over the layout of the phase detectors in both modes. Figure 6.24 shows the phase of the PLL after an SEU on the phase detector. Without TMR, the PLL encountered 235 cycle slips while scanning over the circuit, but only 2 cycle slips were detected while scanning over the TMR phase detector. This measurement shows that there is a big improvement of using the TMR implementation of this circuit.

6.5.4 *CML Divider*

The remaining cross-section that was measured originates from the CML prescaler which was not triplicated. The power consumption in each stage of the divider was 1.5 mW per stage and is implemented as a master–slave flip-flop with CML latches. Especially the CMOS buffers that are located in between the two CML flip-flops are sensitive to SEUs. The phase deviation on the locked PLL will vary depending on the stage of the divider which was hit [157]. Figure 6.25 shows the laser tests of the CML divider. In total there are four CML latches to compose two flip-flops in the /4 prescaler. Various scans were performed on the prescaler at different points in time and those measurements showed different results, so the SEU sensitivity is time

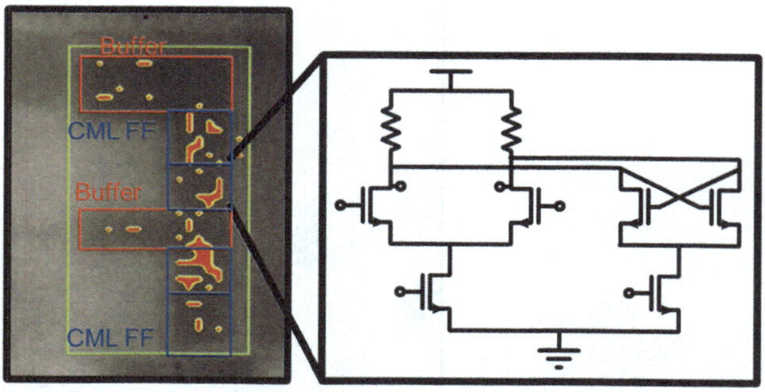

Fig. 6.25 Laser test of the CML divider and the corresponding schematics of a CML latch

dependent. This can be explained by the logic and buffers which are more sensitive during switching. Charges were also injected in the TMR divide-by-16 counter that is implemented with standard CMOS cells. No events were registered by the TDC. To reduce the sensitivity of the PLL, a fully TMR divide-by-64 has been used in an updated circuit with custom standard cells. In this way, single errors on the divider do not contribute to phase jumps in the PLL.

6.6 Practical Implementation

A picture of the SEU experiments at the Cyclotron facility is shown in Fig. 6.26a. The chips and test boards are placed in a vacuum chamber. Due to possible heating issues, the FPGA is placed outside the vacuum chamber and connected with SMA cables to the PLL clocks. The FPGA would require water cooling in the vacuum chamber since no air cooling can occur.

Figure 6.26c shows the experimental setup done at the Cern X-ray facility. X-rays with a spot size of $\sim$2 cm are used to conduct the TID measurement. Figure 6.26b shows a picture of the TPA laser tests. This experiment requires some attention due to the mechanical construction of this test. The problem with TPA lasers is that they cannot pass metal layers in the chip. They need to irradiate the chip from the back side (substrate). This is fine for flip-chip but becomes difficult for wire bonded chips. The approach in this work is shown in Fig. 6.26d. The laser is a conical beam which focuses at the active regions of the chip. To ensure the beam is not interfered by any material (which would cause diffraction patterns), a conical hole is drilled from the back side of the PCB which carries the chip. Sufficient carrier material is still required at the edges of the chip to ensure it can be glued to the PCB and it does not break during air/land transport to the facility. A picture of the board is shown in Fig. 6.26e.

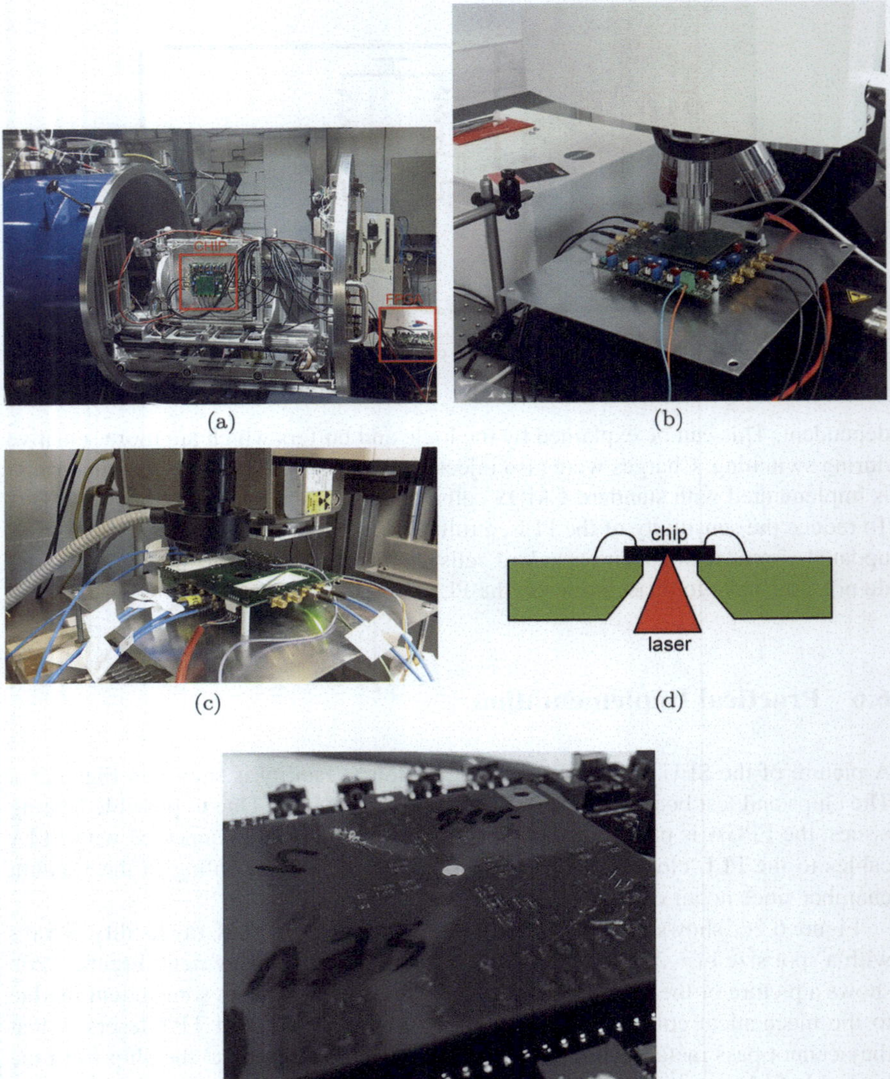

Fig. 6.26 Pictures of the experimental test setup. (**a**) Heavy ion test setup. (**b**) TPA test setup. (**c**) X-ray test setup. (**d**) Cross-section of the PCB used for TPA experiments. (**e**) Backside PCB hole for the TPA laser experiment

6.7 Conclusion

Two PLLs are compared in terms of radiation hardness that contain a ring or LC-oscillator. TID irradiation tests have been done up to 600 Mrad (SiO_2) and two chips were irradiated for comparison. The noise of both PLLs does not change significantly for doses up to 300 Mrad but does change at higher doses due to a change in the bandwidth of the PLL. The ring oscillator shows severe degeneration of 10% to 30% while the LC-oscillator only shows a 5% variation in oscillation frequency after irradiation up to 600 Mrad. SEU experiments are performed with heavy-ions and a Two-Photon Absorption laser experiment with a phase detection TDC embedded in an FPGA. The heavy-ion tests have shown that the LC-oscillator has a larger cross-section than the ring oscillator. The laser tests have identified that the varactor in the LC-oscillator is the most sensitive part of the circuit. Furthermore, these tests showed the various sensitive parts of the PLLs.

From this circuit, the LC-tank oscillator has better TID hardness but has much higher cross-section compared to the ring oscillator. Though for low jitter clock generators, LC-oscillators are preferred due to their excellent phase noise. The next chapter will introduce a solution to mitigate the SEU sensitivity in low noise LC oscillators with the pros and cons for these circuit architectures.

Chapter 7
Radiation Hard Frequency Synthesizers

7.1 Introduction

The discussions from the radiation experiments lead to a conclusion that the varactor of a conventional LC-oscillator is prone to Single-Event Effects due to the large collection area. These effects caused frequency steps since this charge moves to the loop filter of the PLL. In this chapter, an improved circuit is proposed which does not have this drawback. The idea is to avoid the charge collection in the varactors by biasing them in another way. The circuit topology is discussed, followed by heavy-ion experiments which are confirmed with TPA laser experiments.

7.2 Proposed Oscillator Architecture

The original VCO is shown in Fig. 7.1a which was optimized for phase noise [102] and tuning range. The output nodes Vop and Von have a common mode voltage of $V_{dd}/2$ to $V_{dd}/4$. Therefore, the voltage across the varactor, which is DC coupled to the tank, can be tuned from approximately $-V_{dd}/2$ up to $+V_{dd}/2$ (depending on the common mode voltage). This maximizes the tuning range of the oscillator. However, as discussed in previous chapter, this tuning strategy has a major drawback in terms of SEU sensitivity due to the varactor which is shown again in Fig. 7.2. This nwell is a reverse biased junction between the nwell and the substrate and can introduce SETs on the control voltage of the oscillator. The area of this junction is equal

J. Prinzie et al., *Radiation Hardened CMOS Integrated Circuits for Time-Based Signal Processing*, Analog Circuits and Signal Processing,
https://doi.org/10.1007/978-3-319-78616-2_7

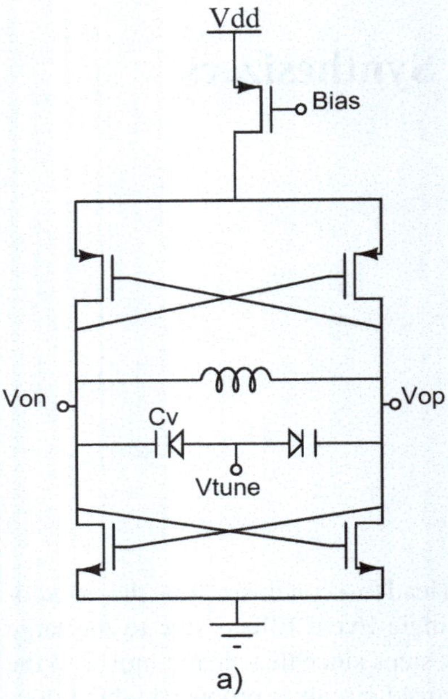

Fig. 7.1 (**a**) Conventional VCO tuning, (**b**) AC-coupled VCO tuning

Fig. 7.2 Varactor cross section

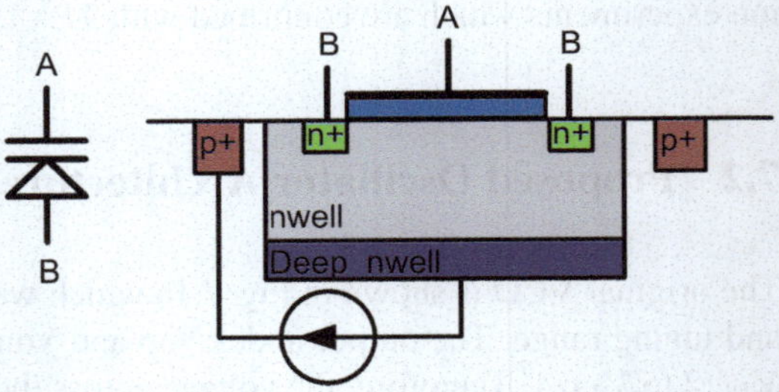

to the area of the varactor which is sensitive to ionizing radiation. Furthermore, a vertical junction is present at the edge of the nwell which introduces a larger charge collection. Since the varactor area of a high-performance oscillator rises above 2500 μm^2, this effect becomes a critical concern. The charge which is collected in the varactor is proportional to the reverse bias voltage of the nwell, the higher this voltage, the more charge is collected.

An improved circuit that mitigates this issue is shown in Fig. 7.1b. The same varactor is used as in the original circuit but it is AC-coupled to the tank through a coupling capacitor C_c and the nwell is shorted to ground. The biasing of the varactor is done with a poly resistor R_b which is connected to the control voltage of the oscillator. This tuning topology provides a significant reduction of the cross section of the PLL since the varactor is not contributing anymore to any SET effects: since the nwell of the varactor is connected to the p-substrate, any charge collected in the nwell to p-substrate junction is shorted through the ground metal wires.

The addition of the coupling capacitor C_c and biasing resistor R_b introduces some design constraints to the circuit. The coupling capacitor not only has a capacitor from the output node to the varactor but also has a bottom-plate capacitance of about 1/10th of C_c parallel to the varactor. In this design, the ratio of C_c to Cvmax has been chosen to be 5 to minimize the effect of the bottom plate capacitance with an affordable coupling factor. Furthermore, the biasing resistor acts as a lossy device parallel to the tank leading to a direct reduction of the quality factor Q of the tank which increases the power consumption and noise of the oscillator. By increasing the value of R_b, the reduction of Q is lowered but the noise due to FM modulation on the varactors generated by R_b increases. The overall FM phase noise contribution of R_b is

$$PN = 4kTR_b \frac{K_{vco}^2}{4\pi^2 f_{off}^2}. \tag{7.1}$$

A 600 Ω resistor was included that results in a 10% Q reduction. Moreover, the additional phase noise was only 2 dBc/Hz compared to a noiseless resistor. Since poly-resistors are used, the tuning node of the varactor is insensitive to ionizing radiation.

A disadvantage of this topology is the tuning curve. The tuning topology used in Fig. 7.1a has the varactors biased at $V_{dd}/2$ on one node. This results in a relatively linear tuning curve. The second topology has the varactor biased at ground on one node resulting in a less linear tuning curve. The result of this is that the gain of the VCO is less constant over the tuning range resulting in non-constant PLL loop dynamics for different VCO control voltages. This effect can be minimized by adding a coarse-fine control loop or discretely tuned capacitors to the tank which are set digitally. This reduces the required range of the control voltage resulting in a more constant VCO gain.

The AC coupling used in the tuning voltage leads to a significant advantage in power supply sensitivity. If the traditional circuit is considered, the gain factor for the supply noise to the common mode output voltage equals

$$A_{supply} = \frac{\frac{1}{2g_{mN}}}{r_{ds} + \frac{1}{2g_{mN}} + \frac{1}{2g_{mP}}}. \tag{7.2}$$

The output phase noise due to the supply noise can be calculated with the gain of the VCO since the varactors are DC connected to the common mode voltage of the oscillator. Therefore, the noise transfer function from the supply noise to the output phase noise is

$$\frac{PN(\Delta f)}{v_{supply}^2} = A_{supply}^2 \frac{Kvco^2}{4\pi^2 \Delta f^2}. \tag{7.3}$$

In the proposed circuits, the term A_{supply} will be identical but in this case, the varactors are not connected directly to the common mode voltages of the tank and the FM conversion of the power supply noise can be neglected at low frequencies.

7.3 Measurements

7.3.1 *Tuning Bands*

The circuit was processed in the same 65 nm CMOS technology and was designed for the same oscillation frequency as the previous design. The measurement of the free running oscillation frequency of both tuning topologies is shown in Fig. 7.3. The tuning range of the original topology is 3.5% while the tuning range of the optimized tuning for radiation effects has been reduced to 3%.

Figure 7.3 also shows the gain of the VCO, which is the derivative of the tuning curve. The varactor biased to ground shows a higher variation of the VCO gain. Typically, in second order PLLs, the damping factor of the loop improves as the gain increases, therefore there will be a maximum control voltage that can be set by the PLL to ensure a stable loop. For this, an additional array of 6 capacitors is added parallel to the varactors to increase the tuning range with sufficient overlap such that the necessary range of the control voltage is reduced. The measurements of the frequency for the optimized tuning strategy are indicated in Fig. 7.4 which shows the overlap of the different tuning bands. If a smaller variation of VCO gain is required, the overlap of the different digitally controlled bands can be increased by adding more, but smaller digitally controlled capacitors to the tank.

7.3.2 *Nominal Tuning Voltage*

It is clear from Fig. 7.4 that the maximum gain of the VCO occurs when the control voltage is zero. However, this bias is not preferred. Typically, in a PLL, the charge pumps prefer to have a voltage at the output node at half of the supply voltage. At this point, the output impedance of the charge pump is maximal and the current imbalance between the up- and down-currents is minimal.

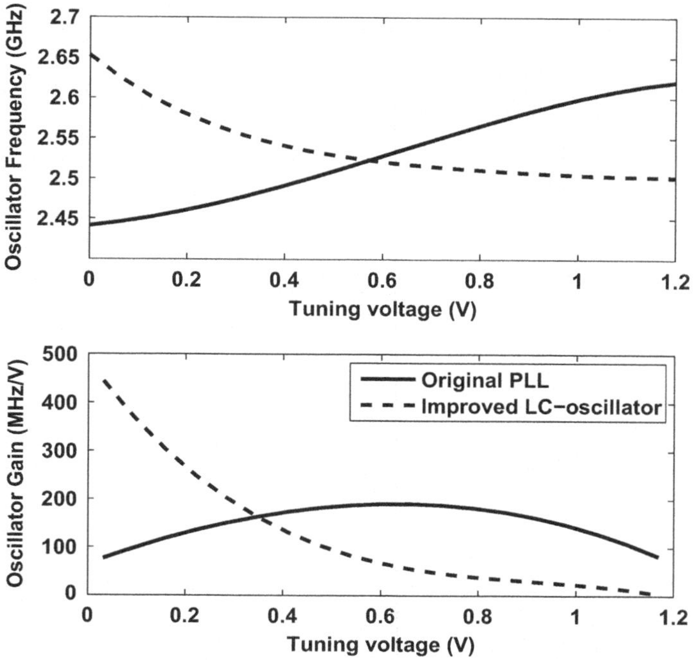

Fig. 7.3 Measurements of the free running oscillation frequency and VCO gain of both oscillators

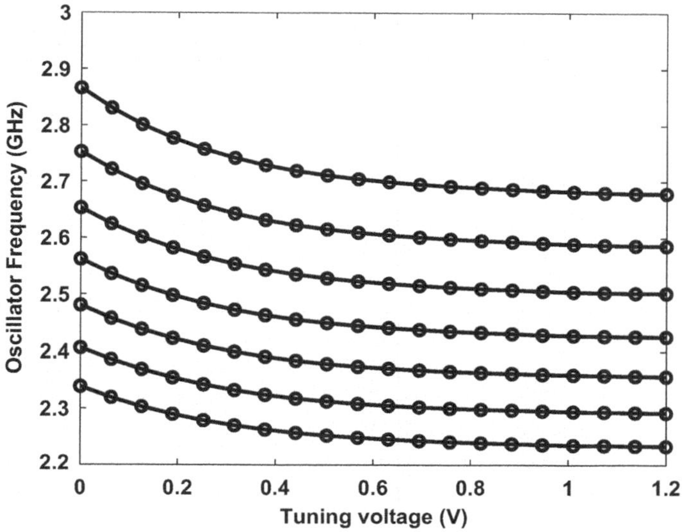

Fig. 7.4 Measurements of the tuning curve for various digitally controlled bands of the optimized circuit

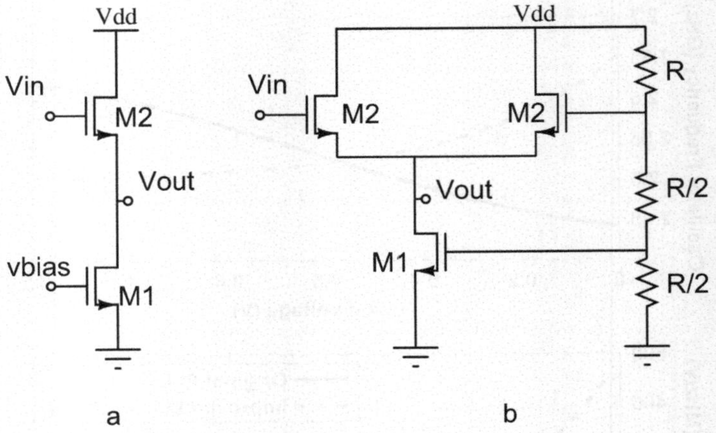

Fig. 7.5 Schematic of the level shifter

If the VCO is typically biased at $V_{dd}/2$, the required range at the charge pump output node is relatively high compared to the case when the VCO is biased at a lower voltage since the tuning curve is much steeper and the overlap between the bands is higher. In that case, a smaller voltage range is required at the output of the charge pump.

To ensure a nominal voltage at the charge pump of $V_{dd}/2$ and minimal voltage range required to tune the VCO at the output of the charge pump, a level shifter is proposed to be inserted between the charge pump's control voltage and the tuning voltage of the VCO. In this design a level shifting of 0.3 V has been implemented to allow a 0.6 V nominal charge pump voltage to be converted to a 0.3 V voltage at the VCO's tuning voltage.

Figure 7.5a shows a basic schematic to implement a level shifter which is relatively PVT independent. Both M1 and M2 are zero-Vt devices and have equal sizes. This ensures that the output voltage equals

$$V_{out} = V_{in} - V_{bias} \tag{7.4}$$

since V_{gs} of both devices is equal. The generation of V_{bias} may lead to some design difficulties especially to ensure high power supply rejection at the output.

Figure 7.5b shows a detailed schematic of the level shifter implemented in this design. In this case, M1 is double the size of M2 but carefully matched in the layout to ensure that $g_{m1} = 2g_{m2}$. At a 1.2 V power supply. The DC voltage generated at the gate of M1 equals 0.3 V and the gate voltage of M2 has a nominal voltage of 0.6 V. In this way, the AC gain of the level shifter is 0.5 with a voltage drop of 0.3 V. However, the right branch of the circuit cancels the noise originating from the supply since 1/2 of the supply noise arrives at the gate of M2 while 1/4th of

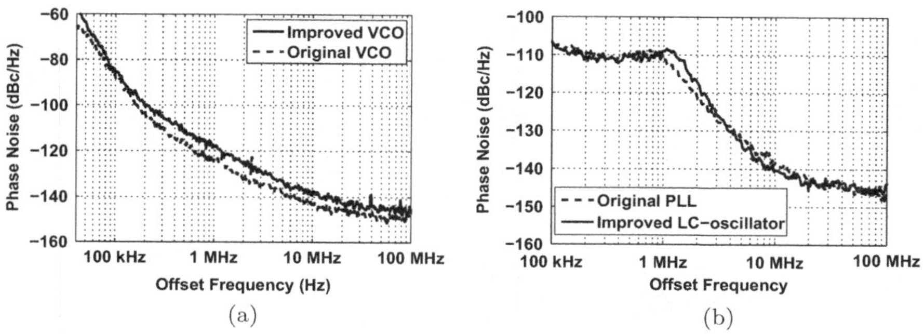

Fig. 7.6 Phase noise measurements. (**a**) Measurements of the open loop VCO phase noise of the original and improved circuit. (**b**) Measurements of the closed loop PLL phase noise of the original and improved circuit

the supply noise arrives at the gate of M1. Since the g_m is double and the gain is inverted, the supply noise cancels at the output. Simulations have shown that the worst case PSRR of this circuit is 22 dB. The overall contribution can be calculated by taking the PLL loop bandwidth into account.

7.3.3 Phase Noise

The phase noise of the improved oscillator was measured to be −118 dBc/Hz at an offset frequency of 1 MHz with a power consumption of 1.8 mW which results in a FOM of 188.7 dB. The original configuration had a comparable FOM of 185 dB. The phase noise measurements are shown in Fig. 7.6a. At high offset frequencies, where $1/f^2$ noise dominates, the additional phase noise is a result from the biasing resistor and a slightly larger inductor which was chosen to reduce the power consumption of the oscillator. However, at low frequencies, the $1/f^3$ noise is similar in both architectures. The excess phase noise increases the overall rms jitter from 350 fs rms measured in the traditional architecture up to 400 fs rms measured in the improved architecture which leads to a negligible increase of overall timing uncertainty in the PLL. The closed loop PLL measurements are shown in Fig. 7.6b.

7.4 Radiation Experiments

Again, two experiments were performed to verify the optimized radiation performance of the VCO compared to the traditional tuning strategy. An irradiation was performed with heavy ions that have an LET from 3.3 up to 62.5 MeV cm^2/mg. The circuit was irradiated with a particle flux of 1.5×10^4 ions/s. The total fluence depended on LET since the experimental time was limited. For our experiment, an LET of 10.5 MeV cm^2/mg was the most interesting value since this equivalent energy transfer is encountered in the LHC experiments such as ATLAS and CMS at CERN.

7.4.1 Heavy-Ion Experiment

The same measurement setup was used as in the first heavy-ion experiment. The fluences measured for each sample (circuit 1 and 2) are shown in Table 7.1.

The cross sections are measured by dividing the SEU count by the fluence and is shown in Fig. 7.7. The measurements of 3 circuits are included in this plot. The first is the traditional LC oscillator. The second is the ring oscillator,

Table 7.1 Ion fluence per LET

LET	Fluence chip 1	Fluence chip 2
62.5	10^7	10^7
32.4	8×10^6	3×10^7
20.4	8×10^6	1.48×10^8
10	8×10^6	10^8
3.3	8×10^6	10^7

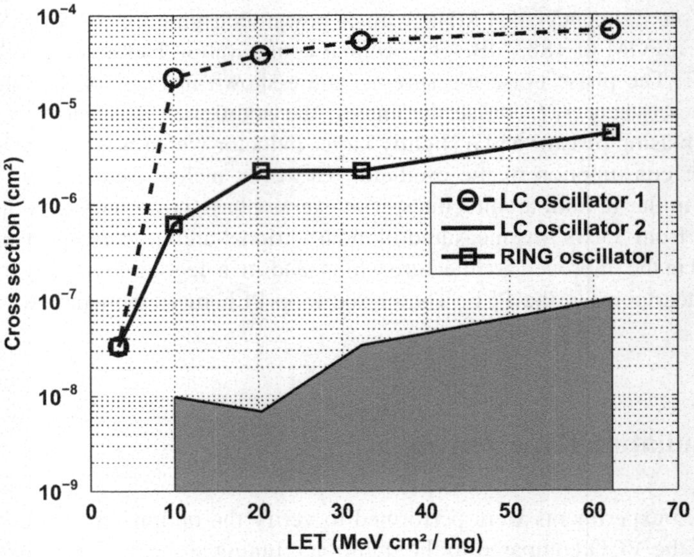

Fig. 7.7 Circuit cross section from heavy ion experiments

as discussed in previous chapter PLL. Third is the optimized LC-oscillator which showed more than two orders of magnitude reduction in cross section. During the experiment, only 1 error was observed at both 62.5 and 32.4 MeV cm^2/mg. At or below 20.4 MeV cm^2/mg, no phase jumps were observed and a value of 1 was used to calculate the maximum estimated cross section. Therefore, the cross section is represented as a shaded area such that it represents the worst case cross section. The optimized oscillator shows an improvement of approximately 800 compared to the traditional varactor tuning. The fluence that was used to test the optimized VCO is higher for LETs of 10 and 20.4 MeV cm^2/mg since these LETs are experimentally more interesting for our applications. The corresponding estimated cross section is therefore smaller for these LETs (since no errors were observed).

7.4.2 *Two Photon Laser Experiment*

A two photon laser experiment was also performed to verify the heavy ion SEU tests. In this laser test, the active area of the VCO was scanned with a step size of 0.2 μm. As discussed in the previous chapter, the traditional VCO has a significant sensitivity to ionizing particles that impinge on the varactors. This cross section was measured with the heavy ion tests and it was verified with the two photon absorption laser test. The optimized VCO proposed in this work did not show any sensitivity while scanning the varactor devices. Different laser energies were utilized from 250 pJ up to 2 nJ which is the maximum energy that can be deposited before destroying the device. The charge was injected 5 μm below the oxide–silicon interface in the substrate which had shown the largest sensitivity before. This confirms that the structure proposed in this work is much more robust against single-event effects.

7.5 Conclusion

An optimized VCO tuning architecture was presented for ionizing radiation environments. The LC-tank resonance frequency is adjusted by an AC-coupled varactor with nwell junctions connected to the ground node that prevents any current injection from high-energy particles into the oscillator or PLL loop filter. An experimental verification was done with high energy particles showing that the cross-section has decreased by more than two orders of magnitude. Two photon laser experiments identified that a traditional varactor tuning topology is more sensitive to ionizing radiation by scanning the active blocks of the oscillator. Noise

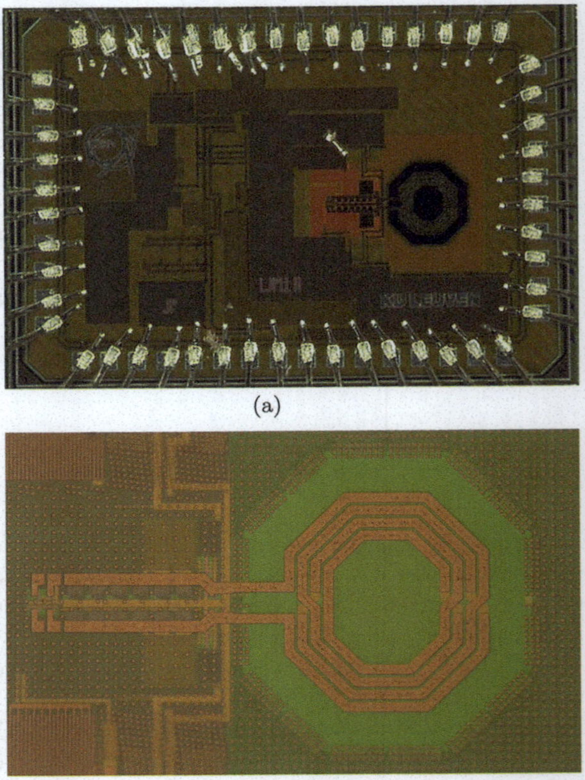

(a)

(b)

Fig. 7.8 Die photographs. (**a**) Die photograph of the PLL. (**b**) Die photograph of the improved oscillator

measurements showed that no significant performance loss as well as tuning range reduction is observed in the improved structure. An array of digitally controlled capacitors is mandatory to increase the frequency range. Overlap between the tuning bands reduces the deterministic variation of VCO gain upon the tuning voltage. A die photograph of the prototype chip is shown in Fig. 7.8a, b.

Chapter 8
Conclusion

In this work, the design aspects of radiation hardened CMOS time-based signal processing circuits were discussed. To gain more insight in this field, different aspects were discussed, starting with the fundamental physics which are involved in the interaction of radiation with silicon, to the practical design implementations of the circuits. In order to validate the proposed circuits and radiation hardening techniques, a detailed experimental study was done to verify the circuits before, during, and after irradiation. These results were used to further reduce the sensitivity of CMOS time-based circuits to ionizing radiation.

In the next section, a conclusion summarizes the findings in this work. Starting from these, an evaluation is made on the industrial valorization of this research. Finally, a suggestion is given for the future work that can still be done in this field.

8.1 Main Conclusions

Radiation hardened CMOS integrated circuits are desired in high-energy physics experiments, nuclear systems, and space applications. Compared to commercial devices, radiation hardened circuits are robust against nuclear radiation. This work presented time-based signal processing circuits which are required for these applications. The main focus were Time-to-Digital converters and clock generators. TDCs are interesting to precisely measure time-intervals and are used to identify particles in high-energy particle detectors or LIDAR Time-of-Flight distance measurements. They are accompanied with low noise clock generators which can filter jitter on the reference clock and can provide a high-frequency clock source based on a low frequency reference. When these circuits are implemented in standard CMOS technologies, however, they tend to suffer from radiation effects degrading their performance. Radiation-hardening by design techniques have been used to mitigate these effects.

J. Prinzie et al., *Radiation Hardened CMOS Integrated Circuits for Time-Based Signal Processing*, Analog Circuits and Signal Processing,
https://doi.org/10.1007/978-3-319-78616-2_8

In Chap. 1, a brief introduction was given to the different mechanisms of high-energy radiation on how it interacts with matter, especially Si and SiO_2. The mechanisms of charged particles, neutrons and photons were discussed. Basically, the energy deposition in the chip results in a generation of free charge carriers. A first effect, on a long time scale, is the so-called Total Ionizing Dose effect which results from an accumulation of trapped charges in the oxides near the transistors. From these discussions, it can be concluded that in modern processes, the effects are dominant from within the STI and LDD extensions. Furthermore, an important observation is that pmos transistors are much more sensitive than nmos transistors. A second effect happens on a very short time scale. Single-Event Effects are generated from single particles which deposit energy in the silicon (substrate). The generated electron-hole pairs are mobile and generate transient currents which can lead to errors in the circuits. These effects were known to be more severe in scaled CMOS technologies. It can be concluded that TID effects are simulated as a separate environmental corner. SEE effects have become more difficult to simulate and involve TCAD simulations and detailed knowledge of the process technology for accurate results. However, approximate and conceptual SPICE simulations can be done using double exponential current sources with a predetermined charge.

An overview of the different design aspects and architectures of modern TDCs was given in Chap. 2. It was shown that TDCs can benefit from CMOS scaling as they operate in the time domain. Reducing gate delays improves the performance of these systems since these circuits are based on delay lines. The TDCs' performance indicators such as linearity and single-shot precision were discussed. Among the different architectures, a Delay Locked Loop has proven beneficial for the stability of the timing resolution in response to TID effects which is catastrophic in the different TDC architectures.

The fundamentals of frequency synthesizers were introduced in Chap. 3 with the frequency domain responses of charge-pump PLLs including a discussion on the different circuit implementations and their non-ideal effects. The fundamentals of oscillators with two methods to calculate the phase noise were discussed. The first method was a linear model which gives a good insight in the design trade-offs for the circuits but which cannot identify the time-variance of the noise in a VCO. The second model proposes an Impulse Sensitivity Function that is a time-dependent impulse response to noise in an oscillator. This model is also useful to study single-event effects in oscillators. The relation between jitter and phase noise was clarified since time-domain circuits prefer jitter as a measure compared to a frequency domain phase noise spectrum. Finally, the different noise sources in a PLL are discussed which can be decomposed in the noise originating from the reference clock, charge-pump, loop filter resistor, VCO and jitter in the digital blocks.

A radiation tolerant TDC is presented in Chap. 4. Its architecture is based on a DLL which ensures that the resolution of the TDC remains constant when TID effects degrade the transistors. The overhead of the DLL implies that it can be disturbed by single-event transients in the circuit. A hybrid phase detector was proposed which combines a bang-bang phase detector with a linear phase-frequency detector. The bang-bang phase detector has a low static phase offset and implies a trade-off between limit cycle jitter and acquisition speed. To ensure a sufficiently small jitter, a PFD was added to improve the startup dynamics of the DLL. To ensure that both phase detectors are not competing, the PFD is automatically disabled through a built-in deadzone. The circuit locks within a 250 ns period with this technique, independently of the initial phase difference. The bang-bang phase detector was studied in detail and a new circuit architecture was proposed to reduce the timing mismatch in bang-bang phase detectors. It uses a two-state architecture in which the offsets in the detector are calibrated in one phase and it operates as a detector in the second phase. The TDC was manufactured in a 40 nm CMOS technology and a resolution of 4.8 ps was achieved with a power consumption of 4.2 mW and a 0.43 LSB rms and 0.4 LSB rms DNL and INL, respectively. The self-calibrating phase detection circuit reduces the static offsets down to 500 fs which is three times better than an uncalibrated circuit.

In the second part of this research, a radiation hardened clock generator was designed for a 64-channel TDC SoC, in collaboration with CERN. The motivation for this research was the study of radiation effects and a comparison between LC-tank oscillators and ring oscillators. In Chap. 5, the different effects of radiation on charge-pump PLLs have been discussed from which it can be concluded that charge-pumps and oscillators are more difficult to fully harden compared to the digital blocks (Divider and PFD). The digital circuits can be implemented with TMR to harden them against single-event effects. However, VCOs continue to suffer from TID effects and single-event effects. It has been proven that LC-tank oscillators are less sensitive to single-event upsets compared to ring oscillators. The circuits which are used in an experimental test chip are discussed. The circuits were manufactured in a 65 nm CMOS technology and were tested for the industrial temperature range. A ring oscillator and an LC-tank oscillator were implemented to verify the performance and radiation hardness for both circuits. The ring-oscillator has a tuning range from 1.5 GHz up to 3.2 GHz and the measured closed loop PLL jitter was 5.2 ps rms. However, temperature experiments from −25 to 125 °C showed a significant dependency of the free running oscillation frequency. The low-noise LC-tank oscillator can be tuned from 2.2 to 3.2 GHz with a typical phase noise of −124 dBc/Hz @ 1 MHz and has a FOM of 185 dBc/Hz with low temperature dependence. The PLL has a configurable bandwidth from 0.7 to 2 MHz. The measured jitter of the full circuit is only 345 fs rms at a power consumption of 11.7 mW resulting in a FOM of −238.5 dB.

The radiation experiments (both TID and single-events) were discussed in Chap. 6. The chip has both a ring oscillator and an LC-tank oscillator. The TID experiments were done with X-rays up to 600 Mrad and showed that the ring oscillator degraded significantly (up to 40%). This implies that the maximum frequency in the VCO reduces due to the reduced currents in the transistors. Furthermore, the gain of the VCO reduces, which leads to an unstable PLL loop. On the other hand, the LC oscillator does not show a significant response at high doses. The free running oscillation frequency remains stable. A similar response was also observed in the phase noise of the closed loop PLLs. Single-event experiments were done using heavy-ions and were verified with Two-Photon Laser experiments. In contrast to the TID effects, the LC-oscillator showed a much higher cross section than the ring oscillator. This was due to the varactor tuning topology which was used in the LC-oscillator. The varactors in an n-well were prone to capture charges which were injected on the loop filter and caused the PLL to lose lock.

In Chap. 7, an improved LC-tank oscillator was presented with an architecture, which mitigates the issues originating in the varactors. The varactors were coupled to the tank using two coupling capacitors which enabled the n-well terminal of the varactor to be connected to the ground. This avoids the collection of charges. The tank is then tuned with a biasing resistor on the varactors' gates. The circuit has a competing performance compared to state-of-the-art circuits which are not radiation hardened. It has been experimentally verified using heavy ions and TPA experiments that this topology is basically insensitive to single-event in the varactors. No errors were observed in these experiments so it can be concluded that the proposed architecture significantly improves the radiation hardness of LC-tank oscillators.

8.2 Major Contributions

The main contributions of this work can be summarized as follows:

- A hybrid phase detector is presented which provides a fast locking mechanism in DLL based TDCs. The circuit improves the recovery time of the DLL after a single-event on the circuit [67].
- A novel bang-bang phase detector is proposed which reduces the static phase offsets and timing mismatch between the circuits [132].
- A low jitter, radiation hardened PLL was designed for a TDC in the high-energy physics community [93, 158].
- A comparison is made between ring oscillators and LC-tank oscillators in terms of TID and single-event sensitivity [159].
- An improved radiation hard LC-tank oscillator was designed and verified experimentally.
- A radiation hardened 2.56 Gbps CDR–5.12 GHz PLL is designed for a 10.24 Gbps serial communication chipset in the high-energy physics community.

8.3 Industrial Valorization

8.3.1 TDCs

The TDC which has been discussed in Chap. 4 has a strong potential to be used in the high-energy physics community, medical imaging applications, and space technologies. Radiation hardened electronics provide a cheaper alternative to high amounts of shielding. Below, a few applications are listed with the necessary developments that are still required.

- **High-energy particle detectors**
  The high-energy physics community has expressed their interest in radiation hardened TDCs with resolutions below 10 ps. These circuits are required to measure time intervals or time-over-threshold of the signals, generated by the particle detectors. The high resolution that was achieved in this research will improve the measurement accuracy of these systems. Its radiation hardness will provide a higher reliability of the readout system such that the chances of information loss are reduced. Although only one-channel TDC was designed in this work, a multi-channel TDC is preferred in most experiments. The future application developments for this kind of circuits will therefore require a focus on a flexible, low power TDC with multiple channels where aspects such as parasitic channel coupling need to be carefully investigated.
- **LIDAR in space applications**
  Distance measurements are intensively performed in space systems. LIDAR is used to observe the height of satellites above a planet's surface. This information is used to map geographical information of the surface. Since the highest elevations on earth approximate 9 km, a dynamic range of at least 54 μs is required. The TDC which has been designed in this work should therefore be extended with a counter to achieve a larger dynamic range. LIDAR altitude measurements may not always require a mm-accuracy, on the other side, inter-satellite ranging has gained interest these days. Originally, docking between two spacecrafts was done manually but laser guided positioning systems have improved the safety and efficiency of the docking maneuver. Some extraterrestrial experiments even require multiple satellites which are positioned accurately to each other and require a moderate (sub-cm) accuracy at several kilometers away from each other. These concepts will definitely benefit from the developments that are done in this research. The radiation hardness of the reported TDC will reduce the required shielding of the electronics that significantly reduces the mission costs and volumetric sizes of the payload.
- **High resolution medical imaging systems**
  As discussed in Chap. 2, TDCs are highly used in PET scanners to measure the arrival times of photons that originate from within a patient. The accuracy of the TDCs improves the performance of these imaging systems and the time-interval measurements require no calibration since its resolution does not degrade due to accumulated doses or environmental variations (such as temperature). This makes the setup of these imaging systems much more easy and cost-efficient.

The space industry has expressed its interest in a commercial circuit which can be used as a stand-alone, high-resolution time-interval measurement device. The prototype chip in this work requires a high speed clock which cannot be generated from a crystal oscillator. A PLL is required for this which has to be radiation tolerant in these harsh environments. Therefore, it is mandatory to integrate a radiation tolerant PLL on the TDC. A prototype which is combining both circuits is now being produced that includes the radiation hardened PLL from this work. **This project has been advanced in collaboration with a company to commercialize the techniques which were discussed in this work.** A TDC is now being developed which achieves a resolution below 10 ps with a 23-bit dynamic range and a flexible serial interface to the chip.

8.3.2 Clock Generators

Major steps have been taken in this work to make high-performance PLLs ready for industrial applications in harsh radiation environments. This has led to many design challenges to achieve the targeted circuit performance within the expected environmental conditions (supply voltages, temperature, radiation) across the different process corners. This work has succeeded to produce a test chip which has been verified to work across the specified voltage and temperature ranges and meeting the performance for all process corners. The PLL uses an automatic frequency calibration to automatically select the correct band of the oscillator which may vary due to process variations.

- **Clock generators for high-resolution TDCs**
  Throughout this dissertation, it has been explained that DLL based TDCs require a high frequency reference clock. This necessitates a reference clock with a low accumulated jitter during the measured time-interval of the TDC. For space or high-energy physics experiments, these clock generators have to be radiation tolerant. The protection against TID effects (which has been tested up to 600 Mrad) ensures that the PLL can operate for a long duration (up to 10 years) in the worst environments of the LHC.
- **High-speed serial communication links**
  PLLs are the fundamental building blocks of high-speed serial communication links. In these circuits, a PLL is used in two ways. Firstly, the PLL can generate the clock signal that drives a high speed serializer which transmits the data on an electrical or optical link. Secondly, the PLL can work as a clock recovery circuit to recover the clock signal from a data stream. The PLL is the most critical block which determines the performance of such data communication chips. It is essential that the PLL's performance remains constant after being irradiated up to the expected doses. This PLL yields an excellent performance across the total doses that will be expected in the High-Luminosity LHC (HL-LHC). Furthermore, it is crucial that the PLL is insensitive to single-event effects to prevent any loss of data during transmission. Even if error correction is used, a single-event in the PLL can result in a loss of synchronization which means that the link goes down and some data may be lost.
- **Low noise clock generators for space applications**
  In earth's orbital satellites or deep space probes, the modules require reliable clock signals which are the heartbeats of many systems. Single-event transients

in these clocks may be disastrous in the logic, processors and synchronization devices. Since deep space probes have a long mission duration, its accumulated dose can also become too large for commercially available, space-grade components. However, this PLL can work up to extremely high doses, which is sufficient for these missions that can last several decades. The circuit may need additional verification. This includes certified radiation experiments, shock, and vacuum tests to be certified as a space grade product.

- **Frequency synthesizers for radiation tolerant wireless links**
Wireless communication has gained more interests in nuclear facilities to reduce the mass of electrical or optical transmission media. The costs and volumetric loss of these cables can be overcome with wireless links. In such links, direct conversion transceivers are based on a low noise frequency synthesizer to upconvert the digital data to RF or mm-wave frequencies. For products that can be used in a wide range of applications, the division ratio has to be made programmable or even include a fractional-N divider. The tuning range of the output may also need to be improved to cover a range from f_1 up to $2f_1$ such that all frequencies can be generated through a division.

The PLL in this work has been highly valorized at the ATLAS, CMS, and other experiment at CERN. **The PLL is used in two commercial chips that are put in production in 2018**. The PLL is used in the following two chips which are being developed at CERN:

- **A radiation hardened optical transceiver project at CERN**
The "Low Power Gigabit Transceiver" (LPGBT) is a 10.24 Gbps serial communication link which acts as a data aggregation module that combines multiple low-speed electrical links into a fast optical link between the detectors and the computing rooms. Besides the findings in this work, also a radiation hardened CDR has been designed for the LPGBT, based on the experimental conclusions that were discussed here. The circuits are planned for submission in fall 2017 and will be mass-produced in 2018. The LPGBT is the communication module which is used by **all** the detectors of the CERN experiments to transmit the data between the detectors (and sub-detectors) to the central computing rooms. The radiation requirements for this chipset are targeted to reach a total dose of 200 Mrad and will be installed in the LHC detectors for the phase 2 upgrades during the Long Shutdown 3, around 2023.
- **A 64-channel TDC with 3 ps resolution project at CERN**
The "PICOTDC" is a high-resolution TDC with 3 ps resolution and 64 time-interval channels which has been introduced in Chap. 5. This chip is a revolutionary TDC with a flexible architecture with extensive data buffering and different readout interfaces. A low performance (25 ps) system is now being used in more than 20 high-energy physics applications (ALICE TOF, CMS muon detectors, STAR, BES, KABES,HADES, NICA, NA62, AMS, Belle, BES, etc.). This chip will replace this system with up to 20,000 devices. The PLL from this research provides a stable, reliable, and low noise clock for the TDC. Its robustness against

high doses and single-event upsets ensures that the critical timing module of this SoC is extremely reliable.

The nuclear and space industry has expressed its interest for the development of a flexible, radiation hard clock generator which can generate a broad range of frequencies, based on a low frequency reference clock. This device can then be used for clock synthesis and jitter reduction in various applications, like nuclear power plants, satellites, cube-sats, and particle physics experiments. A commercialization of such a device is now being considered.

8.4 Future Work

8.4.1 TDCs

High resolution TDCs, used in industrial applications provide endless possibilities. However, some concerns still apply and only a few topics are covered within this work. Process variations result in a significant variation of the delay of the VCDL. Compared to a typical process corner, the fast corner running at a 10% higher supply voltage will have more than twice the speed compared to the typical corner. The contrary is true for a slow process corner at a lower supply voltage. This implies some serious constraints on the VCDL's tuning range. The architecture presented in this research requires that the total delay of the VCDL can be tuned to match the reference clock period in the DLL. This tuning range, taking into account the variations, is difficult to achieve with a normal delay line topology, especially if a military operating temperature range is required. A solution is to use multiple delay lines which have a slightly different average delay. In this way, with some overlap in the tuning curves, a wider range can be covered. During an in-factory test, the correct delay line can be selected or a state machine, comparable to the PLL, can digitally select the correct line when the chip is powered on. Another approach is split the chips into two or more operating speed grades after an in-factory test.

Nevertheless, some fields may demand even better resolutions for a time-interval measurement. The challenge is not only to acquire sub-picosecond accuracies but also to achieve a good linearity and single-shot precision for such a resolution. Many ultra-high resolution TDCs rely on time amplifiers which are intrinsically prone to variations. These circuits may be an interesting field of research.

In interpolating delay lines (like the one discussed here), the resolution may become limited due to the mismatch between the different interpolated stages of the timing generator, especially if a sub-picosecond resolution is targeted. To achieve this precision, an on-chip self calibration is required which can tune the individual delay cells. Such a system may rely on a proper reference calibration scheme and individually (digitally) tunable delay cells.

Furthermore, as discussed in Chap. 2, TDCs benefit from CMOS scaling and advanced processes such as sub-10 nm finfet technologies which can boost the performance of the TDCs but the effect of radiation for low-jitter applications in such technologies surely presents a further topic of research.

Single-event effects on the DLL were mitigated in this research, however, some attention may be required on the delay lines. A single-event (e.g., an additional pulse in the line) may not disturb the DLL but may result in an incorrect time-interval measurement. TMR is a straightforward but a power/area hungry solution which can be optimized, based on the knowledge of the periodic signals in the VCDL. Furthermore, SOI (Silicon on Insulator) technologies can provide an improved single-event tolerance since they are intrinsically less sensitive to collect charges, compared to bulk CMOS technologies. However, they may only be useful in short-period space applications where the total dose is not high since SOI (both FDSOI (Fully Depleted SOI) and PDSOI (Partially Depleted SOI)) suffer from large TID effects due to the Burried Oxide (BOX) below the transistors. Compensating for these effects calls for additional research.

Finally, all digital TDC implementations which can be fully synthesized are interesting to reduce the design overhead and the sensitivity of the analog blocks to radiation. Furthermore, these techniques will be beneficial for FPGA implementations. Stochastic TDCs are good candidates for such highly digital TDCs for harsh environments which can be widely explored in future research.

8.4.2 Clock Generators

The clock synthesizer from this research was designed for a frequency of only 2.56 GHz and, in a later stage, a 5.12 GHz oscillator for the LPGBT project. However, larger frequencies may be required in the future. For example, in mm-wave communications or communication links up to 100 Gbps, a higher clock frequency may be required. This implies that the capacitance in the tank is reduced which makes such high-speed oscillators more sensitive to single-event upsets. These high-speed circuits will require much more developments for operation in harsh radiation environments. An alternative is to use a moderate frequency range ($<$10 GHz) and to use multiple phases to enhance the speed of the circuit.

The next steps in the research of radiation hardened clock generators may need to focus on all-digital PLLs. From the discussions in this work, it can be concluded that the charge-pumps cannot be hardened completely, though the effects can be made acceptable. In all-digital PLLs, the only analog circuit that remains vulnerable is the oscillator. All other analog functions of the PLL can be implemented in the digital domain which can be easily hardened against single-event upsets by using TMR. To reduce the power consumption, not all blocks may require TMR which can be an interesting field of research.

From this work, frequency synthesizers are well understood in how they react to ionizing radiation. However, if a fully wireless radiation hardened transceiver is targeted, much more work is required to mitigate radiation effects on the circuit, or architectural level of the transceiver. Not only the frequency synthesizer can be prone to radiation but also the LNAs, PAs, mixers, and ADCs suffer from radiation effects.

From the first experiments of the LC-tank oscillators, it was clear that the varactor was very sensitive to ionizing radiation. In fact, even a PLL could be used to detect, or amplify and convert charges to the phase domain. These results may then be read by phase- or frequency deviations rather than voltage transients. This leads to the domain of frequency based sensor interfaces which have gained interest in low voltage scaled technologies.

8.4.3 Radiation Testing

Finally, during this research, a significant amount of work was done to experimentally test the circuits in radiation facilities. The most challenging issue that was encountered was the correlation between the single-event effects from the heavy-ions and the two-photon laser tests. From the heavy-ion tests, a cross section can be estimated and the LET from these particles is well-known. However, during the laser test, one has control over the laser energy but this does not directly correlate to the generated charges in the silicon. The amounts of charges which are generated highly depend on substrate doping. Therefore, it may be difficult, with limited knowledge of the process technology, to estimate which laser energy does approximately result in a similar charge injection from a heavy-ion. Since heavy-ion tests are extremely expensive and time-consuming, it will be both cost and time efficient if these tests can be replaced with a laser experiment.

Bibliography

1. Michael G. Stabin, editor. *Interaction of Radiation with Matter*. Springer New York, New York, NY, 2007.
2. Brian R. Martin. *Nuclear and Particle Physics: An Introduction, 2nd Edition*. Wiley, 2009.
3. Ervin B. Podgorsak. *Radiation Physics for Medical Physicists*. Springer, 2010.
4. Friedel Weinert Daniel Greenberger, Klaus Hentschel. *Compendium of Quantum Physics*. Springer, 2009.
5. James E. Martin. *Physics for Radiation Protection: A Handbook, Second Edition*. Wiley, 2006.
6. R. C. Lacoe. Improving integrated circuit performance through the application of hardness-by-design methodology. *IEEE Transactions on Nuclear Science*, 55(4):1903–1925, Aug 2008.
7. J. R. Schwank, M. R. Shaneyfelt, D. M. Fleetwood, J. A. Felix, P. E. Dodd, P. Paillet, and V. Ferlet-Cavrois. Radiation effects in MOS oxides. *IEEE Transactions on Nuclear Science*, 55(4):1833–1853, Aug 2008.
8. R. C. Lacoe, J. V. Osborn, R. Koga, S. Brown, and D. C. Mayer. Application of hardness-by-design methodology to radiation-tolerant ASIC technologies. *IEEE Transactions on Nuclear Science*, 47(6):2334–2341, Dec 2000.
9. H. J. Barnaby. Total-ionizing-dose effects in modern CMOS technologies. *IEEE Transactions on Nuclear Science*, 53(6):3103–3121, Dec 2006.
10. N. S. Saks, D. B. Brown, and R. W. Rendell. Effects of switched gate bias on radiation-induced interface trap formation [MOS transistors]. *IEEE Transactions on Nuclear Science*, 38(6):1130–1139, Dec 1991.
11. S. N. Rashkeev, D. M. Fleetwood, R. D. Schrimpf, and S. T. Pantelides. Effects of hydrogen motion on interface trap formation and annealing. *IEEE Transactions on Nuclear Science*, 51(6):3158–3165, Dec 2004.
12. Kwok K. Ng Simon M. Sze. *Physics of Semiconductor Devices, 3rd Edition*. Wiley, 2006.
13. L. Ratti, L. Gaioni, M. Manghisoni, G. Traversi, and D. Pantano. Investigating degradation mechanisms in 130 nm and 90 nm commercial CMOS technologies exposed to up to 100 mrad ionizing radiation dose. In *2007 9th European Conference on Radiation and Its Effects on Components and Systems*, pages 1–9, Sept 2007.
14. S. Gerardin, M. Bagatin, D. Cornale, L. Ding, S. Mattiazzo, A. Paccagnella, F. Faccio, and S. Michelis. Enhancement of transistor-to-transistor variability due to total dose effects in 65-nm MOSFETs. *IEEE Transactions on Nuclear Science*, 62(6):2398–2403, Dec 2015.

J. Prinzie et al., *Radiation Hardened CMOS Integrated Circuits for Time-Based Signal Processing*, Analog Circuits and Signal Processing,
https://doi.org/10.1007/978-3-319-78616-2

15. A. Scarpa, A. Paccagnella, F. Montera, G. Ghibaudo, G. Pananakakis, G. Ghidini, and P. G. Fuochi. Ionizing radiation induced leakage current on ultra-thin gate oxides. *IEEE Transactions on Nuclear Science*, 44(6):1818–1825, Dec 1997.
16. F. Faccio, S. Michelis, D. Cornale, A. Paccagnella, and S. Gerardin. Radiation-induced short channel (RISCE) and narrow channel (RINCE) effects in 65 and 130 nm MOSFETs. *IEEE Transactions on Nuclear Science*, 62(6):2933–2940, Dec 2015.
17. F. Faccio and G. Cervelli. Radiation-induced edge effects in deep submicron CMOS transistors. *IEEE Transactions on Nuclear Science*, 52(6):2413–2420, Dec 2005.
18. A. H. Johnston, R. T. Swimm, G. R. Allen, and T. F. Miyahira. Total dose effects in CMOS trench isolation regions. *IEEE Transactions on Nuclear Science*, 56(4):1941–1949, Aug 2009.
19. Lihua Dai, Xiaonian Liu, Mengying Zhang, Leqing Zhang, Zhiyuan Hu, Dawei Bi, Zhengxuan Zhang, and Shichang Zou. Degradation induced by TID radiation and hot-carrier stress in 130-nm short channel PDSOI NMOSFETs. *Microelectronics Reliability*, 74:74–80, 2017.
20. M. Silvestri, S. Gerardin, A. Paccagnella, and F. Faccio. Degradation induced by x-ray irradiation and channel hot carrier stresses in 130-nm NMOSFETs with enclosed layout. *IEEE Transactions on Nuclear Science*, 55(6):3216–3223, Dec 2008.
21. M. Silvestri, S. Gerardin, A. Paccagnella, F. Faccio, and L. Gonella. Channel hot carrier stress on irradiated 130-nm NMOSFETs. *IEEE Transactions on Nuclear Science*, 55(4):1960–1967, Aug 2008.
22. G. Anelli, M. Campbell, M. Delmastro, F. Faccio, S. Floria, A. Giraldo, E. Heijne, P. Jarron, K. Kloukinas, A. Marchioro, P. Moreira, and W. Snoeys. Radiation tolerant vlsi circuits in standard deep submicron CMOS technologies for the LHC experiments: practical design aspects. *IEEE Transactions on Nuclear Science*, 46(6):1690–1696, Dec 1999.
23. W. J. Snoeys, T. A. P. Gutierrez, and G. Anelli. A new NMOS layout structure for radiation tolerance. *IEEE Transactions on Nuclear Science*, 49(4):1829–1833, Aug 2002.
24. Miryala S. Kulis S. Christiansen J. Francisco R. Casas L.M.J., Ceresal D. and Gnani D. Characterization of radiation effects in 65nm digital circuits with the DRAD digital radiation test chip. In *Topical workshop on electronics for Particle Physics (TWEPP),*. IEEE, 26–30 Sept. 2016.
25. P. E. Dodd. Device simulation of charge collection and single-event upset. *IEEE Transactions on Nuclear Science*, 43(2):561–575, Apr 1996.
26. D. A. Black, W. H. Robinson, I. Z. Wilcox, D. B. Limbrick, and J. D. Black. Modeling of single event transients with dual double-exponential current sources: Implications for logic cell characterization. *IEEE Transactions on Nuclear Science*, 62(4):1540–1549, Aug 2015.
27. M. Mitrović, M. Hofbauer, B. Goll, K. Schneider-Hornstein, R. Swoboda, B. Steindl, K. O. Voss, and H. Zimmermann. Experimental investigation of single-event transient waveforms depending on transistor spacing and charge sharing in 65-nm CMOS. *IEEE Transactions on Nuclear Science*, 64(8):2136–2143, Aug 2017.
28. B. Narasimham, B. L. Bhuva, R. D. Schrimpf, L. W. Massengill, M. J. Gadlage, O. A. Amusan, W. T. Holman, A. F. Witulski, W. H. Robinson, J. D. Black, J. M. Benedetto, and P. H. Eaton. Characterization of digital single event transient pulse-widths in 130-nm and 90-nm CMOS technologies. *IEEE Transactions on Nuclear Science*, 54(6):2506–2511, Dec 2007.
29. P. E. Dodd, M. R. Shaneyfelt, J. A. Felix, and J. R. Schwank. Production and propagation of single-event transients in high-speed digital logic ICs. *IEEE Transactions on Nuclear Science*, 51(6):3278–3284, Dec 2004.
30. P. E. Dodd and L. W. Massengill. Basic mechanisms and modeling of single-event upset in digital microelectronics. *IEEE Transactions on Nuclear Science*, 50(3):583–602, June 2003.
31. R. García Alía, M. Brugger, S. Danzeca, V. Ferlet-Cavrois, C. Frost, R. Gaillard, J. Mekki, F. Saigné, A. Thornton, S. Uznanski, and F. Wrobel. SEL hardness assurance in a mixed radiation field. *IEEE Transactions on Nuclear Science*, 62(6):2555–2562, Dec 2015.
32. F. W. Sexton. Destructive single-event effects in semiconductor devices and ICs. *IEEE Transactions on Nuclear Science*, 50(3):603–621, June 2003.

33. P. E. Dodd, M. R. Shaneyfelt, J. R. Schwank, and J. A. Felix. Current and future challenges in radiation effects on CMOS electronics. *IEEE Transactions on Nuclear Science*, 57(4): 1747–1763, Aug 2010.
34. R. Terada and M. Watanabe. Error injection analysis for triple modular and penta-modular redundancies. In *2017 6th International Symposium on Next Generation Electronics (ISNE)*, pages 1–4, May 2017.
35. H. B. Wang, Y. Q. Li, L. Chen, L. X. Li, R. Liu, S. Baeg, N. Mahatme, B. L. Bhuva, S. J. Wen, R. Wong, and R. Fung. An SEU-tolerant dice latch design with feedback transistors. *IEEE Transactions on Nuclear Science*, 62(2):548–554, April 2015.
36. J. E. Knudsen and L. T. Clark. An area and power efficient radiation hardened by design flip-flop. *IEEE Transactions on Nuclear Science*, 53(6):3392–3399, Dec 2006.
37. S. Kulis. Single event effects mitigation with TMRG tool. *Journal of Instrumentation*, 12(01):C01082, 2017.
38. R. W. Blaine, N. M. Atkinson, J. S. Kauppila, S. E. Armstrong, N. C. Hooten, T. D. Loveless, J. H. Warner, W. T. Holman, and L. W. Massengill. Differential charge cancellation (DCC) layout as an RHBD technique for bulk CMOS differential circuit design. *IEEE Transactions on Nuclear Science*, 59(6):2867–2871, Dec 2012.
39. O. A. Amusan, A. F. Witulski, L. W. Massengill, B. L. Bhuva, P. R. Fleming, M. L. Alles, A. L. Sternberg, J. D. Black, and R. D. Schrimpf. Charge collection and charge sharing in a 130 nm CMOS technology. *IEEE Transactions on Nuclear Science*, 53(6):3253–3258, Dec 2006.
40. J. S. Kauppila, A. L. Sternberg, M. L. Alles, A. M. Francis, J. Holmes, O. A. Amusan, and L. W. Massengill. A bias-dependent single-event compact model implemented into BSIM4 and a 90 nm CMOS process design kit. *IEEE Transactions on Nuclear Science*, 56(6):3152–3157, Dec 2009.
41. J. Allison et al. Geant4 developments and applications. *IEEE Transactions on Nuclear Science*, 53(1):270–278, Feb 2006.
42. M. Raine, M. Gaillardin, P. Paillet, and O. Duhamel. Towards a generic representation of heavy ion tracks to be used in engineering see simulation tools. *IEEE Transactions on Nuclear Science*, 61(4):1791–1798, Aug 2014.
43. M. Alioto. Ultra-low power vlsi circuit design demystified and explained: A tutorial. *IEEE Transactions on Circuits and Systems I: Regular Papers*, 59(1):3–29, Jan 2012.
44. P. R. Kinget. Scaling analog circuits into deep nanoscale CMOS: Obstacles and ways to overcome them. In *2015 IEEE Custom Integrated Circuits Conference (CICC)*, pages 1–8, Sept 2015.
45. X. Jiang, X. Jiang, P. Malcovati, V. Stojanovic, and T. Lizuka. F1: Digitally assisted analog and analog-assisted digital in high-performance scaled CMOS process. In *2014 IEEE International Solid-State Circuits Conference Digest of Technical Papers (ISSCC)*, pages 510–511, Feb 2014.
46. J. P. Jansson, P. Keränen, J. Kostamovaara, and A. Baschirotto. CMOS technology scaling advantages in time domain signal processing. In *2017 IEEE International Instrumentation and Measurement Technology Conference (I2MTC)*, pages 1–5, May 2017.
47. Stephan Henzler. *Time-to-Digital Converters*. Springer, 2010.
48. T. J. Yamaguchi, S. Komatsu, M. Abbas, K. Asada, N. N. Mai-Khanh, and J. Tandon. A CMOS flash TDC with 0.84–1.3 ps resolution using standard cells. In *2012 IEEE Radio Frequency Integrated Circuits Symposium*, pages 527–530, June 2012.
49. C. F. Bedoya, J. Marin, J. C. Oller, and C. Willmott. Electronics for the CMS muon drift tube chambers: The read-out minicrate. *IEEE Transactions on Nuclear Science*, 52(4):944–949, Aug 2005.
50. G. Cerminara. The drift tube system of the CMS experiment. *Nuclear Physics B - Proceedings Supplements*, 172:71–74, 2007. Proceedings of the 10th Topical Seminar on Innovative Particle and Radiation Detectors.
51. X. Llopart, M. Campbell, R. Dinapoli, D. San Segundo, and E. Pernigotti. Medipix2: A 64-k pixel readout chip with 55- mu;m square elements working in single photon counting mode. *IEEE Transactions on Nuclear Science*, 49(5):2279–2283, Oct 2002.

52. L. Blanquart, J. Richardson, K. Einsweiler, P. Fischer, E. Mandelli, G. Meddeler, and I. Peric. Fe-i2: a front-end readout chip designed in a commercial 0.25- mu;m process for the atlas pixel detector at LHC. *IEEE Transactions on Nuclear Science*, 51(4):1358–1364, Aug 2004.
53. V. Y. Panin, F. Kehren, C. Michel, and M. Casey. Fully 3-d pet reconstruction with system matrix derived from point source measurements. *IEEE Transactions on Medical Imaging*, 25(7):907–921, July 2006.
54. J. Langner. Development of a Parallel Computing Optimized Head Movement Correction Method in Positron Emission Tomography. Master of computer science thesis, University of Applied Sciences Dresden and Forschungszentrum Dresden-Rossendorf, 2003.
55. M. G. Strauss, R. Brenner, F. J. Lynch, and C. B. Morgan. 2-d position-sensitive scintillation detector for neutrons. *IEEE Transactions on Nuclear Science*, 28(1):800–806, Feb 1981.
56. C. Niclass, M. Soga, H. Matsubara, M. Ogawa, and M. Kagami. A 0.18-mu m CMOS SoC for a 100-m-range 10-frame/s 200 , $times$,96-pixel time-of-flight depth sensor. *IEEE Journal of Solid-State Circuits*, 49(1):315–330, Jan 2014.
57. J. P. Jansson, V. Koskinen, A. Mantyniemi, and J. Kostamovaara. A multichannel high-precision CMOS time-to-digital converter for laser-scanner-based perception systems. *IEEE Transactions on Instrumentation and Measurement*, 61(9):2581–2590, Sept 2012.
58. Y. Cao, W. De Cock, M. Steyaert, and P. Leroux. Design and assessment of a 6 ps-resolution time-to-digital converter with 5 MGy gamma-dose tolerance for LIDAR application. *IEEE Transactions on Nuclear Science*, 59(4):1382–1389, Aug 2012.
59. R. B. Staszewski, J. L. Wallberg, S. Rezeq, Chih-Ming Hung, O. E. Eliezer, S. K. Vemulapalli, C. Fernando, K. Maggio, R. Staszewski, N. Barton, Meng-Chang Lee, P. Cruise, M. Entezari, K. Muhammad, and D. Leipold. All-digital PLL and transmitter for mobile phones. *IEEE Journal of Solid-State Circuits*, 40(12):2469–2482, Dec 2005.
60. Wu Gao, Deyuan Gao, Christine Hu Guo, and Yann Hu. Integrated high-resolution multi-channel time-to-digital converters (TDCs) for pet imaging, 01 2011.
61. T. Enami, S. Ninomiya, and M. Hashimoto. Statistical timing analysis considering spatially and temporally correlated dynamic power supply noise. *IEEE Transactions on Computer-Aided Design of Integrated Circuits and Systems*, 28(4):541–553, April 2009.
62. T. Na, S. H. Woo, J. Kim, H. Jeong, and S. O. Jung. Comparative study of various latch-type sense amplifiers. *IEEE Transactions on Very Large Scale Integration (VLSI) Systems*, 22(2):425–429, Feb 2014.
63. J. A. McNeill. Jitter in ring oscillators. *IEEE Journal of Solid-State Circuits*, 32(6):870–879, Jun 1997.
64. P. Dudek, S. Szczepanski, and J. V. Hatfield. A high-resolution CMOS time-to-digital converter utilizing a vernier delay line. *IEEE Journal of Solid-State Circuits*, 35(2):240–247, Feb 2000.
65. J. Yu, F. F. Dai, and R. C. Jaeger. A 12-bit vernier ring time-to-digital converter in 0.13 $muhboxm$ CMOS technology. *IEEE Journal of Solid-State Circuits*, 45(4):830–842, April 2010.
66. A. Liscidini, L. Vercesi, and R. Castello. Time to digital converter based on a 2-dimensions vernier architecture. In *2009 IEEE Custom Integrated Circuits Conference*, pages 45–48, Sept 2009.
67. J. Prinzie, M. Steyaert, and P. Leroux. A single shot TDC with 4.8 ps resolution in 40 nm CMOS for high energy physics applications. *Journal of Instrumentation*, 10(01):C01031, 2015.
68. S. Henzler, S. Koeppe, W. Kamp, H. Mulatz, and D. Schmitt-Landsiedel. 90nm 4.7ps-resolution 0.7-LSB single-shot precision and 19pj-per-shot local passive interpolation time-to-digital converter with on-chip characterization. In *2008 IEEE International Solid-State Circuits Conference - Digest of Technical Papers*, pages 548–635, Feb 2008.
69. C. Yao, F. Jonsson, J. Chen, and L. R. Zheng. A high-resolution time-to-digital converter based on parallel delay elements. In *2012 IEEE International Symposium on Circuits and Systems*, pages 3158–3161, May 2012.

70. A. Samarah and A. C. Carusone. A digital phase-locked loop with calibrated coarse and stochastic fine TDC. *IEEE Journal of Solid-State Circuits*, 48(8):1829–1841, Aug 2013.
71. J. P. Jansson, A. Mantyniemi, and J. Kostamovaara. Synchronization in a multilevel CMOS time-to-digital converter. *IEEE Transactions on Circuits and Systems I: Regular Papers*, 56(8):1622–1634, Aug 2009.
72. Y. H. Seo, J. S. Kim, H. J. Park, and J. Y. Sim. A 0.63ps resolution, 11b pipeline TDC in 0.13 um CMOS. In *2011 Symposium on VLSI Circuits - Digest of Technical Papers*, pages 152–153, June 2011.
73. A. Elkholy, T. Anand, W. S. Choi, A. Elshazly, and P. K. Hanumolu. A 3.7 mw low-noise wide-bandwidth 4.5 GHz digital fractional-n PLL using time amplifier-based TDC. *IEEE Journal of Solid-State Circuits*, 50(4):867–881, April 2015.
74. J. Yu, F. F. Dai, and R. C. Jaeger. A 12-bit vernier ring time-to-digital converter in 0.13 $muhboxm$ CMOS technology. *IEEE Journal of Solid-State Circuits*, 45(4):830–842, April 2010.
75. Dietrich Schlichthärle. *Oversampling and Noise Shaping*, pages 469–491. Springer Berlin Heidelberg, Berlin, Heidelberg, 2011.
76. M. Z. Straayer and M. H. Perrott. A multi-path gated ring oscillator TDC with first-order noise shaping. *IEEE Journal of Solid-State Circuits*, 44(4):1089–1098, April 2009.
77. K. Okuno, T. Konishi, S. Izumi, M. Yoshimoto, and H. Kawaguchi. A 62-db SNDR second-order gated ring oscillator TDC with two-stage dynamic d-type flipflops as a quantization noise propagator. In *10th IEEE International NEWCAS Conference*, pages 289–292, June 2012.
78. Y. Cao, W. De Cock, M. Steyaert, and P. Leroux. 1-1-1 mash $deltasigma$ time-to-digital converters with 6 ps resolution and third-order noise-shaping. *IEEE Journal of Solid-State Circuits*, 47(9):2093–2106, Sept 2012.
79. S. Ito, S. Nishimura, H. Kobayashi, S. Uemori, Y. Tan, N. Takai, T. J. Yamaguchi, and K. Niitsu. Stochastic TDC architecture with self-calibration. In *2010 IEEE Asia Pacific Conference on Circuits and Systems*, pages 1027–1030, Dec 2010.
80. Xilinx 7 series FPGAs configurable logic block : https://www.xilinx.com/support/documentation/user_guides/ug474_7Series_CLB.pdf.
81. S. Y. Wang, J. Wu, S. H. Yao, and W. C. Chang. A field-programmable gate array (FPGA) TDC for the Fermilab Seaquest (e906) experiment and its test with a novel external wave union launcher. *IEEE Transactions on Nuclear Science*, 61(6):3592–3598, Dec 2014.
82. Bram De Muer and Michiel Steyaert. *CMOS Fractional-N Synthesizers*. Springer US, 2003.
83. J. W. Moon, K. C. Choi, and W. Y. Choi. A 0.4-v, 90 to 350-MHz PLL with an active loop-filter charge pump. *IEEE Transactions on Circuits and Systems II: Express Briefs*, 61(5):319–323, May 2014.
84. Behzad Razavi. *RF Microelectronics (2Nd Edition) (Prentice Hall Communications Engineering and Emerging Technologies Series)*. Prentice Hall Press, Upper Saddle River, NJ, USA, 2nd edition, 2011.
85. S. K. Garg and B. Singh. A novel design of an efficient low power phase frequency detector for delay locked loop. In *2016 IEEE 1st International Conference on Power Electronics, Intelligent Control and Energy Systems (ICPEICES)*, pages 1–4, July 2016.
86. Behzad Razavi. *Designing BangBang PLLs for Clock and Data Recovery in Serial Data Transmission Systems*, pages 34–45. Wiley-IEEE Press, 2003.
87. Jri Lee, K. S. Kundert, and B. Razavi. Analysis and modeling of bang-bang clock and data recovery circuits. *IEEE Journal of Solid-State Circuits*, 39(9):1571–1580, Sept 2004.
88. C. Hangmann, I. Wüllner, C. Hedayat, and U. Hilleringmann. Modeling and characterization of CP-PLL phase noise in presence of dead zone. In *2014 IEEE 12th International New Circuits and Systems Conference (NEWCAS)*, pages 349–352, June 2014.
89. A. Samarah and A. C. Carusone. Cycle-slipping pull-in range of bang-bang PLLs. In *2015 IEEE 13th International New Circuits and Systems Conference (NEWCAS)*, pages 1–4, June 2015.

90. V. S. Sadeghi, H. Miar Naimi, and M. P. Kennedy. The role of charge pump mismatch in the generation of integer boundary spurs in fractional-n frequency synthesizers: Why worse can be better. *IEEE Transactions on Circuits and Systems II: Express Briefs*, 60(12):862–866, Dec 2013.
91. H. Shao, K. Lin, B. Wang, C. Chen, F. Gao, F. Huang, and X. Wang. A high-performance charge pump with improved static and dynamic matching characteristic. In *2015 IEEE 11th International Conference on ASIC (ASICON)*, pages 1–4, Nov 2015.
92. C. Zhang, T. Au, and M. Syrzycki. A high performance NMOS-switch high swing cascode charge pump for phase-locked loops. In *2012 IEEE 55th International Midwest Symposium on Circuits and Systems (MWSCAS)*, pages 554–557, Aug 2012.
93. J. Prinzie, J. Christiaensen, P. Moreira, M. Steyaert, and P. Leroux. A low noise clock generator for high-resolution time-to-digital convertors. *Journal of Instrumentation*, 11(02):C02038, 2016.
94. Y. S. Park, D. h. Kwon, K. C. Choi, and W. Y. Choi. Multi-band VCO PLL with continuously maintained optimum VCO control voltage. *Electronics Letters*, 50(12):857–859, June 2014.
95. A. Sheikholeslami. Source degeneration [circuit intuitions]. *IEEE Solid-State Circuits Magazine*, 6(3):5–6, Summer 2014.
96. A. Dyskin and I. Kallfass. Source degeneration as series-series feedback. In *2015 IEEE International Conference on Microwaves, Communications, Antennas and Electronic Systems (COMCAS)*, pages 1–3, Nov 2015.
97. C. T. Charles and D. J. Allstot. A buffered charge pump with zero charge sharing. In *2008 IEEE International Symposium on Circuits and Systems*, pages 2633–2636, May 2008.
98. K. Arshak, O. Abubaker, and E. Jafer. Improved charge pump for reduced clock feed through and charge sharing suppression. In *Proceedings of the Fifth IEEE International Caracas Conference on Devices, Circuits and Systems, 2004.*, volume 1, pages 192–194, Nov 2004.
99. J. Luo, L. Zhang, L. Zhang, Y. Wang, and Z. Yu. Behavioral analysis and optimization of CMOS CML dividers for millimeter-wave applications. *IEEE Transactions on Circuits and Systems II: Express Briefs*, 62(3):256–260, March 2015.
100. Lutz von Wangenheim. On the Barkhausen and Nyquist stability criteria. *Analog Integrated Circuits and Signal Processing*, 66(1):139–141, Jan 2011.
101. M. Tiebout. Low-power low-phase-noise differentially tuned quadrature VCO design in standard CMOS. *IEEE Journal of Solid-State Circuits*, 36(7):1018–1024, Jul 2001.
102. D. Ham and A. Hajimiri. Concepts and methods in optimization of integrated LC VCOs. *IEEE Journal of Solid-State Circuits*, 36(6):896–909, Jun 2001.
103. Behzad Razavi. *A Study of Phase Noise in CMOS Oscillators*, pages 176–188. Wiley-IEEE Press, 2003.
104. S. Cui, B. Banerjee, and V. Acharya. A linearized current-controlled oscillator for ultra-low power wideband and cognitive radios. In *2012 IEEE 55th International Midwest Symposium on Circuits and Systems (MWSCAS)*, pages 146–149, Aug 2012.
105. D. B. Leeson. A simple model of feedback oscillator noise spectrum. *Proceedings of the IEEE*, 54(2):329–330, Feb 1966.
106. A. A. Abidi. Phase noise and jitter in CMOS ring oscillators. *IEEE Journal of Solid-State Circuits*, 41(8):1803–1816, Aug 2006.
107. J. Craninckx and M. Steyaert. Low-noise voltage-controlled oscillators using enhanced LC-tanks. *IEEE Transactions on Circuits and Systems II: Analog and Digital Signal Processing*, 42(12):794–804, Dec 1995.
108. A. Hajimiri and T. H. Lee. Design issues in CMOS differential LC oscillators. *IEEE Journal of Solid-State Circuits*, 34(5):717–724, May 1999.
109. A. Hajimiri and T. H. Lee. A general theory of phase noise in electrical oscillators. *IEEE Journal of Solid-State Circuits*, 33(2):179–194, Feb 1998.
110. Ali Hajimiri and Thomas H Lee. *The Design of Low Noise Oscillators*. Springer US, 1999.
111. A. Hajimiri, S. Limotyrakis, and T. H. Lee. Jitter and phase noise in ring oscillators. *IEEE Journal of Solid-State Circuits*, 34(6):790–804, Jun 1999.

112. K. Kundert. Predicting the phase noise and jitter of PLL-based frequency synthesizers : http://www.designers-guide.org/analysis/PLLnoise+jitter.pdf.
113. Adem Aktas and Mohammed Ismail. *PLL Phase Noise Analysis*, pages 21–40. Springer US, Boston, MA, 2004.
114. J. Phillips and K. Kundert. An introduction to cyclostationary noise. *http://www.designers-guide.org/theory/cyclo-preso.pdf.*
115. X. Gao, E. A. M. Klumperink, P. F. J. Geraedts, and B. Nauta. Jitter analysis and a benchmarking figure-of-merit for phase-locked loops. *IEEE Transactions on Circuits and Systems II: Express Briefs*, 56(2):117–121, Feb 2009.
116. K Klein. The Phase-2 Upgrade of the CMS Tracker. Technical Report CERN-LHCC-2017-009. CMS-TDR-014, CERN, Geneva, Jun 2017.
117. D De Gruttola. Particle identification with the ALICE time-of-flight detector at the LHC. *Journal of Instrumentation*, 9(10):C10019, 2014.
118. Aida Todri, M Turqueti, Ryan Rivera, and Simon Kwan. Power distribution studies for CMS forward tracker. In *IEEE Nuclear Science Symposium Conference Record*, pages 1208–1211, 12 2009.
119. O. Popescu. Power budgets for cubesat radios to support ground communications and inter-satellite links. *IEEE Access*, 5:12618–12625, 2017.
120. J. R. Jenness. Radiative cooling of satellite-borne electronic components. *Proceedings of the IRE*, 48(4):641–643, April 1960.
121. L. Perktold and J. Christiansen. A fine time-resolution 3 ps-rms time-to-digital converter for highly integrated designs. In *2013 IEEE International Instrumentation and Measurement Technology Conference (I2MTC)*, pages 1092–1097, May 2013.
122. C. Ljuslin, J. Christiansen, A. Marchioro, and O. Klingsheim. An integrated 16-channel CMOS time to digital converter. *IEEE Transactions on Nuclear Science*, 41(4):1104–1108, Aug 1994.
123. M. J. E. Lee, W. J. Dally, T. Greer, Hiok-Tiaq Ng, R. Farjad-Rad, J. Poulton, and R. Senthinathan. Jitter transfer characteristics of delay-locked loops - theories and design techniques. *IEEE Journal of Solid-State Circuits*, 38(4):614–621, Apr 2003.
124. G. Marucci, S. Levantino, P. Maffezzoni, and C. Samori. Analysis and design of low-jitter digital bang-bang phase-locked loops. *IEEE Transactions on Circuits and Systems I: Regular Papers*, 61(1):26–36, Jan 2014.
125. R. Nonis, W. Grollitsch, T. Santa, D. Cherniak, and N. D. Dalt. A 2.4psrms-jitter digital PLL with multi-output bang-bang phase detector and phase-interpolator-based fractional-n divider. In *2013 IEEE International Solid-State Circuits Conference Digest of Technical Papers*, pages 356–357, Feb 2013.
126. Y. H. Moon, I. S. Kong, Y. S. Ryu, and J. K. Kang. A 2.2-mw 135-MHz false-lock-free dll for display interface in 0.15- μm CMOS. *IEEE Transactions on Circuits and Systems II: Express Briefs*, 61(8):554–558, Aug 2014.
127. J. Verbeeck, M. Van Uffelen, M. S. J. Steyaert, and P. Leroux. Design of a MGy tolerant instrumentation amplifier using a correlated double sampling technique in 130 nm CMOS. In *2011 12th European Conference on Radiation and Its Effects on Components and Systems*, pages 156–159, Sept 2011.
128. Masaya Miyahara, Yusuke Asada, Daehwa Paik, and Akira Matsuzawa. A low-noise self-calibrating dynamic comparator for high-speed ADCs. In *2008 IEEE Asian Solid-State Circuits Conference*, pages 269–272, Nov 2008.
129. M. Lee and A. A. Abidi. A 9 b, 1.25 ps resolution coarse fine time-to-digital converter in 90 nm CMOS that amplifies a time residue. *IEEE Journal of Solid-State Circuits*, 43(4):769–777, April 2008.
130. R. B. Staszewski, S. Vemulapalli, P. Vallur, J. Wallberg, and P. T. Balsara. 1.3 v 20 ps time-to-digital converter for frequency synthesis in 90-nm CMOS. *IEEE Transactions on Circuits and Systems II: Express Briefs*, 53(3):220–224, March 2006.
131. J. S. Kim, Y. H. Seo, Y. Suh, H. J. Park, and J. Y. Sim. A 300-ms/s, 1.76-ps-resolution, 10-b asynchronous pipelined time-to-digital converter with on-chip digital background calibration in 0.13- um CMOS. *IEEE Journal of Solid-State Circuits*, 48(2):516–526, Feb 2013.

132. J. Prinzie, M. Steyaert, and P. Leroux. A self-calibrated bang bang phase detector for low-offset time signal processing. *IEEE Transactions on Circuits and Systems II: Express Briefs*, 63(5):453–457, May 2016.
133. Ricardo Reis, Yu Cao, and Gilson Wirth. *Circuit Design for Reliability*. Springer-Verlag New York, 2015.
134. T. P. Wang and S. Y. Wang. A low-voltage low-power low-phase-noise wide-tuning-range 0.18- um CMOS VCO with high-performance FOMT of 196.3 dbc/hz. In *2013 IEEE MTT-S International Microwave Symposium Digest (MTT)*, pages 1–4, June 2013.
135. Hongda Zheng, Dongxu Yang, Dajie Zeng, Li Zhang, and Zhiping Yu. Design of a 24GHz wide tuning-range VCO with optimized switches in resonator. In *2010 IEEE International Conference of Electron Devices and Solid-State Circuits (EDSSC)*, pages 1–4, Dec 2010.
136. D. S. Lee, J. H. Jang, H. G. Park, Y. Pu, K. C. Hwang, Y. Yang, M. K. Seo, and K. Y. Lee. A wide-locking-range dual injection-locked frequency divider with an automatic frequency calibration loop in 65-nm CMOS. *IEEE Transactions on Circuits and Systems II: Express Briefs*, 62(4):327–331, April 2015.
137. S. J. Horst, S. D. Phillips, J. D. Cressler, K. Kruckmeyer, R. Eddy, A. Aude, P. O'Farrell, B. Zhang, E. Wilcox, and K. LaBel. A study of total dose mitigation approaches for charge pumps in phase-locked loop applications. *IEEE Transactions on Nuclear Science*, 58(6): 3038–3045, Dec 2011.
138. T. D. Loveless, L. W. Massengill, W. T. Holman, B. L. Bhuva, D. McMorrow, and J. H. Warner. A generalized linear model for single event transient propagation in phase-locked loops. *IEEE Transactions on Nuclear Science*, 57(5):2933–2947, Oct 2010.
139. Z. Chen, M. Lin, Y. Zheng, Z. Wei, S. Huang, and S. Zou. Single-event transient characterization of a radiation-tolerant charge-pump phase-locked loop fabricated in 130 nm PD-SOI technology. *IEEE Transactions on Nuclear Science*, 63(4):2402–2408, Aug 2016.
140. T. D. Loveless, L. W. Massengill, B. L. Bhuva, W. T. Holman, R. A. Reed, D. McMorrow, J. S. Melinger, and P. Jenkins. A single-event-hardened phase-locked loop fabricated in 130 nm CMOS. *IEEE Transactions on Nuclear Science*, 54(6):2012–2020, Dec 2007.
141. V. Prasad and S. Sandya. Single event transient tolerant high speed phase frequency detector for pll based frequency synthesizer. In *International Conference on Circuits, Communication, Control and Computing*, pages 77–80, Nov 2014.
142. S. J. Horst, S. D. Phillips, P. Saha, J. D. Cressler, D. McMorrow, and P. Marshall. A theory of single-event transient response in cross-coupled negative resistance oscillators. *IEEE Transactions on Nuclear Science*, 57(6):3349–3357, Dec 2010.
143. S. M. Jung and J. M. Roveda. A radiation-hardened-by-design phase-locked loop using feedback voltage controlled oscillator. In *Sixteenth International Symposium on Quality Electronic Design*, pages 103–106, March 2015.
144. T. Wang, K. Wang, L. Chen, A. Dinh, B. Bhuva, and R. Shuler. A RHBD LC-tank oscillator design tolerant to single-event transients. *IEEE Transactions on Nuclear Science*, 57(6): 3620–3625, Dec 2010.
145. Y. Boulghassoul, L. W. Massengill, A. L. Sternberg, and B. L. Bhuva. Effects of technology scaling on the set sensitivity of RF CMOS voltage-controlled oscillators. *IEEE Transactions on Nuclear Science*, 52(6):2426–2432, Dec 2005.
146. J. G. Maneatis. Low-jitter process-independent DLL and PLL based on self-biased techniques. *IEEE Journal of Solid-State Circuits*, 31(11):1723–1732, Nov 1996.
147. P. C. Huang, W. S. Chang, and T. C. Lee. 21.2 a 2.3GHz fractional-n dividerless phase-locked loop with -112dbc/hz in-band phase noise. In *2014 IEEE International Solid-State Circuits Conference Digest of Technical Papers (ISSCC)*, pages 362–363, Feb 2014.
148. Pin-En Su and S. Pamarti. A 2-mhz bandwidth fractional-n synthesizer based on a fractional frequency divider with digital spur suppression. In *2010 IEEE Radio Frequency Integrated Circuits Symposium*, pages 413–416, May 2010.
149. D. Tasca, M. Zanuso, G. Marzin, S. Levantino, C. Samori, and A. L. Lacaita. A 2.9 to 4.0-GHz fractional-n digital PLL with bang-bang phase detector and 560-fs rms integrated jitter at 4.5-mw power. *IEEE Journal of Solid-State Circuits*, 46(12):2745–2758, Dec 2011.

150. CERN. http://proj-xraymic.web.cern.ch/proj-xraymic/.
151. P. Andreani and S. Mattisson. On the use of MOS varactors in RF VCOs. *IEEE Journal of Solid-State Circuits*, 35(6):905–910, June 2000.
152. R. L. Bunch and S. Raman. Large-signal analysis of MOS varactors in CMOS -gm LC VCOs. *IEEE Journal of Solid-State Circuits*, 38(8):1325–1332, Aug 2003.
153. M. Menouni, M. Barbero, F. Bompard, S. Bonacini, D. Fougeron, R. Gaglione, A. Rozanov, P. Valerio, and A. Wang. 1-grad total dose evaluation of 65 nm CMOS technology for the HL-LHC upgrades. *Journal of Instrumentation*, 10(05):C05009, 2015.
154. Farshid Samii Amlashi, Adib Abrishamifar, and Reza Ebrahimi Atani. Jitter peaking investigation in charge pump based clock and data recovery circuits. In *AFRICON 2007*, pages 1–6, Sept 2007.
155. D. McMorrow, W. T. Lotshaw, J. S. Melinger, S. Buchner, Y. Boulghassoul, L. W. Massengill, and R. L. Pease. Three-dimensional mapping of single-event effects using two photon absorption. *IEEE Transactions on Nuclear Science*, 50(6):2199–2207, Dec 2003.
156. M. Newton, B. Danger, H. Wang, L. Chen, D. M. Hiemstra, and V. Kirischian. Two photon absorption laser facility for single event effect testing. In *2016 IEEE Radiation Effects Data Workshop (REDW)*, pages 1–5, July 2016.
157. T. D. Loveless, B. D. Olson, B. L. Bhuva, W. T. Holman, C. C. Hafer, and L. W. Massengill. Analysis of single-event transients in integer-*n* frequency dividers and hardness assurance implications for phase-locked loops. *IEEE Transactions on Nuclear Science*, 56(6):3489–3498, Dec 2009.
158. J. Prinzie, M. Steyaert, P. Leroux, J. Christiansen, and P. Moreira. A single-event upset robust, 2.2 GHz to 3.2 GHz, 345 fs jitter PLL with triple-modular redundant phase detector in 65 nm CMOS. In *2016 IEEE Asian Solid-State Circuits Conference (A-SSCC)*, pages 285–288, Nov 2016.
159. J. Prinzie, J. Christiansen, P. Moreira, M. Steyaert, and P. Leroux. Comparison of a 65 nm CMOS ring- and LC-oscillator based PLL in terms of TID and SEU sensitivity. *IEEE Transactions on Nuclear Science*, 64(1):245–252, Jan 2017.

Index

J. Prinzie et al., *Radiation Hardened CMOS Integrated Circuits for Time-Based Signal Processing*, Analog Circuits and Signal Processing,
https://doi.org/10.1007/978-3-319-78616-2

The manufacturer's authorised representative in the EU is Springer Nature Customer Service Centre GmbH, Europaplatz 3, 69115 Heidelberg, Germany. If you have any concerns regarding our products, please contact ProductSafety@springernature.com

Printed and bound by CPI Group (UK) Ltd, Croydon, CR0 4YY
05/10/2026
02228144-0001